T0212710

SpringerBriefs in Environmental Science

More information about this series at http://www.springer.com/series/8868

Sergio O. Saldaña-Zorrilla

Natural Disasters, Foreign Trade and Agriculture in Mexico

Public Policy for Reducing Economic Vulnerability

 Springer

Sergio O. Saldaña-Zorrilla
Research Scholar
Risk, Policy and Vulnerability Program
International Institute for Applied Systems Analysis (IIASA)
Laxenburg, Austria

ISSN 2191-5547   ISSN 2191-5555   (electronic)
SpringerBriefs in Environmental Science
ISBN 978-3-319-17358-0   ISBN 978-3-319-17359-7   (eBook)
DOI 10.1007/978-3-319-17359-7

Library of Congress Control Number: 2015940008

Springer Cham Heidelberg New York Dordrecht London

Printed on acid-free paper

Springer International Publishing AG Switzerland is part of Springer Science+Business Media
(www.springer.com)

Abstract

The increasing frequency and economic losses from natural disasters within the framework of decreasing agricultural prices and trade liberalization is becoming crucial in increasing poverty in the Mexican rural economy. During the past two decades, the governmental withdrawal from supporting the agricultural sector with investments in physical, financial and logistic instruments continues to stress agricultural livelihoods, as current private mechanisms have not replaced them effectively. It has contributed to making the agricultural sector particularly vulnerable to a number of hazards as it has weakened economic agents' response and impeded assets accumulation. This book identifies economic vulnerability to natural and economic hazards in order to assess public and private coping capacity, and provides a conceptual framework and economic theory that supports the overall approach and employed methodologies. It is based on quantitative and qualitative research methods, and makes use of econometric analysis and stakeholders' views aimed at finding feasible solutions. Further, this book offers a spatial model that can support policy-decision-making for the creation of differential investments in productive infrastructure, as well as financial instruments to reduce current vulnerability and poverty throughout the national territory.

During the past three decades, over 80% of total economic losses from weather-related disasters occurred in the agricultural sector. In the same period, mean weighted agricultural prices have decreased over 50% in real terms, and since 1996 a trade deficit has persisted in this sector. Currently, insufficient credit access, low coverage of crop insurance, as well as a near lack of investments to expand irrigation and further productive infrastructure, is sharpening the vulnerability of rural livelihoods. These facts explain why this sector produces only 4% of the GDP despite employing over 20% of the national workforce. These facts undermine farmers' expectations of future incomes within the community, stimulating rural-urban out migration, which usually cannot be absorbed by the urban economy in light of the modest industrial dynamism of recent years. This leads to enlargement of informal sectors in large cities and migratory flows to abroad, among others.

Acknowledgments

The author wishes to thank the following persons and institutions for contributing to the success of this book. Above all, special thanks to Dr. Joanne Linerooth-Bayer, Program Leader of the *Risk and Vulnerability Program* at the *International Institute for Applied Systems Analysis* (IIASA), who continually supported this work. Valuable criticism and suggestions from the following people have been decisive in writing this book: Dr. Reinhard Mechler, Dr. Uwe Schubert, Dr. Krister Sandberg, Dr. Neal Leary, Dr. Larry Willmore and Dr. Jose M. Albala-Bertrand.

Particularly valuable was the encouragement obtained from the *World Bank* and the *Provention Consortium*, as well as from the START-IIASA *Advanced Institute on Vulnerability to Global Environmental Change*, which financed the various country consultations.

Contents

Chapter 1
Introduction

Along with economic and institutional stressors, the occurrence of natural disasters is increasingly determining income and asset accumulation of a vulnerable population in Mexico. This book identifies and estimates the effect of natural hazards, along with some economic variables, on poverty, as well as assesses the impact of current public and private strategies aimed at reducing economic vulnerability. Most of the workforce living in poverty in rural areas relies on agricultural livelihoods and they are increasingly affected by weather-related disasters. Insufficient physical and financial instruments to hedge the agricultural sector both from weather- and market-related risks contribute to the weakening response of farmers, the negative consequences of which expand to other regions.

During the period 1980-2013, over 80% of total damages from weather-related events affected the agricultural sector in Mexico. Though the contribution of the agricultural sector is only around 3% of GDP, it is the livelihood of nearly 20% of the national population. Trade liberalization, in Mexico since the mid-1980s, has not only resulted in a negligible increase in production, but also contributed to an increase of uncertainty with regards to small farm incomes, with a rise in imports and depressed prices of agricultural products.

This book recognizes that natural disasters and trade liberalization are not the only factors contributing to poverty in this country; however, it aims to prove that they are becoming economically crucial in maintaining and increasing structural poverty with their disruptive effect on the asset accumulation process. As confirmed in this book's analysis, it drives small-scale farmers out of business, and increases out migration from rural areas. The purpose of this analysis is to examine how natural disasters and adverse economic conditions affect agricultural livelihoods and the implications of these stressors upon the poor in the agricultural sector, in order to come up with suggestions to reduce their vulnerability. Special emphasis is given to the effects of these hazards on migration patterns. In addition, this project analyzes selected policy options (e.g. government-supported insurance schemes) to reduce the vulnerability of farmers. To further this aim, this work makes use of descriptive and econometric analyses, as well as stakeholders' interviews.

© The Author 2015

S.O. Saldaña-Zorrilla, *Natural Disasters, Foreign Trade and Agriculture in Mexico*, SpringerBriefs in Environmental Science, DOI 10.1007/978-3-319-17359-7_1

During the past three decades, the rising amount of research work, particularly in the natural sciences, has contained warnings about the increasing pressure of natural hazards on human societies. So far, the economic sciences have given little attention to these studies. This book presents an integral analysis of both natural and economic hazards as determinants of poverty and labor mobility, embracing social and policy aspects. The methodology includes both quantitative and qualitative methods. A review of existing economic and policy literature is presented mainly in Chapters 2 and 4, which discuss evidence supporting this book's hypothesis and describe the process behind economic vulnerability in Mexico. Chapter 5 is based on a spatial econometrics analysis of data from the 2,443 municipalities of Mexico as well as on a survey of farmers and further stakeholders. Data was gathered from a number of sources, including national and international organizations, as well as those collected directly from households and policymakers during the survey and country consultations carried out in 2003, 2004, 2005 and 2014. This analysis quantitatively validates the hypotheses presented in this text.

The conceptual framework of this book, presented in **Chapter 2**, is based on the vast existing literature from both the natural and social sciences. It describes the basic mechanisms by which vulnerable societies tend to amplify the negative consequences of natural disasters, given their usual weak coping and adaptive capacity. Coping capacity, as discussed, is in turn conditioned not only by natural hazards and market imperfections, but also a result of the original endowment modified by the historic process of wealth distribution, giving shape to the current entitlement system. In addition, the inherent capacity of economic agents to adapt their structure and functioning in response to harmful events is closely conditioned by existing institutions. Given the prevailing high levels of poverty and vulnerability in Mexico, as discussed in Chapter 2, the current public intervention to strengthen coping and adaptive capacity of the exposed population is being exceeded

Considering the increasing tendency of frequent and severe natural disasters and the economic losses associated with them, together with the dropping agricultural prices presented in **Chapter 3**, this research warns of the most likely forthcoming increase in poverty if no effective strategy has been implemented to reduce vulnerability. As the urban economy has experienced a very modest growth during the last three decades, it is unable to absorb the additional workforce released from rural areas. It has led to the proliferation of slums and to a currently fast growing informal urban economy. As both the uncertain agricultural incomes and the precarious urban jobs do not fulfill the expectation of the vulnerable population, increasing emigration to the USA is the usual response, particularly in the case of farmers living in remote regions. This trend is further verified both by the stakeholders' survey and the spatial model.

Chapter 4 assesses current policy instruments for reducing agricultural vulnerability, identifying areas where public investments in productive infrastructure can greatly achieve progress in poverty reduction if combined with disaster mitigation works and expanding public-private insurance schemes. In addition, it analyzes possible improvements to policies and instruments for reducing economic losses, e.g., subsidized micro-insurance, and for reducing poverty and vulnerability in the agricultural sector. In light of the current comparatively low implementation of ex-ante instruments, this chapter underlines the need for a more active promotion of existing disaster mitigation instruments (e.g. funds for mitigation works) from the federal authority along with more participative action from the community and municipal authorities to propose projects. This chapter's research is based to a large extent on a project the author carried out in 2003 for the World Bank and the Prevention Consortium (Saldaña-Zorrilla 2004), which was complemented by a consultation with policymakers in Mexico between 2004 and 2014, a policy assess-ment report prepared for the United Nations (Saldaña-Zorrilla 2007) and a stake-holders' assessment carried out in 2014.

The great human and geophysical diversity of Mexico is reflected in its existing plurality of economic practices and natural hazards exposure. **Chapter 5** applies spatial econometric analyses to assess the importance of natural disasters and depressed prices due to imports (or to simple low international prices), and analyzes their differential interaction with the diminishing income of vulnerable regions to stimulate out migration from municipalities highly exposed to natural hazards. During the past two decades, failure to implement effective mechanisms for reduc-ing risk to climatic and market variability is impeding accumulating enough physi-cal and human assets in, particularly, the agricultural sector. The negative effects of hazards in vulnerable regions have a spillover effect on neighboring regions. For that reason, direct public intervention is needed to allow asset accumulation. Derived from the migration model, one can observe that a segment more prone to emigrate is one that with greater reductions in real incomes and more frequently affected by disasters. It reveals the role of expectations of future incomes in the configuration of migratory flows.

The current role of the government, the private market (e.g., insurers), the farm-ers and other stakeholders in dealing with adverse outcomes from natural disasters and income uncertainty of small-scale farmers are still insufficiently interconnected, as verified by the views of the stakeholders presented in **Chapter 6**. This chapter identifies the problem of farmer vulnerability to natural disasters and trade liberal-ization, and examines the preferred policy responses from the point of view of stakeholders, including farmers, farmer cooperatives, crop insurers, and ministries and departments of local and national governments. A dramatic governmental with-drawal of the previous two decades from supporting the agricultural sector with facilities for production and commercialization continues putting stress on rural livelihoods as no effective compensatory mechanisms have been successfully implemented. This fieldwork identified a clear need for a more profit-oriented pro-duction and expansion of coverage of individual and collective financing instru-ments to face external shocks. The chapter is derived from research the author conducted between 2004 and 2005 as part of a project for the START-IIASA *Advanced Institute of Vulnerability to Global Environmental Change* (Saldaña-Zorrilla 2006) as well as on field work carried out in 2014 in order to elaborate the 10 years follow-up.

Chapter 7 integrates the results of all chapters to provide some wrapping-up remarks and presents a final discussion to guide future directions.

This book's analyses contribute to the current debate of the long-term economic impact of disasters, as well as offer an integral methodology combining natural and social sciences for further studies of country and community level vulnerability to climate change. The lessons derived from this analysis provide useful elements for the design and improvement of governmental policies concerning social and economic development as well. In addition, the desegregation of this analysis has the advantage of facilitating the design and evaluation of governmental projects at municipal, sub-national and national level, as well as provides conceptual-empirical elements for international cooperation in matters of disaster risk reduction, climate change adaptation, rural development and poverty reduction.

Introduction References

Saldaña-Zorrilla SO (2007) Socio-Economic Vulnerability to Natural Disasters in Mexico: rural poor, trade and public response. CEPAL/ECLAC-Economic Commission for Latin-American and Caribbean Countries. United Nations, Mexico City

Saldaña-Zorrilla SO (2006) Views in Reducing Rural Vulnerability to Natural Disasters in Southern Mexico: Hazard Exposure, Coping and Adaptive Capacity. Working paper of the Advanced Institute of Vulnerability to Global Environmental Change. START-IIASA, Washington, DC

Saldaña-Zorrilla SO (2004) Mexican agriculture and its vulnerability to natural disasters and trade. In: Lessons Learned from Natural Disasters Management in Developing Countries (compendium). The World Bank. Washington, DC

Chapter 2
A conceptual framework of economic vulnerability

... Una historia que toma sentido a partir de la conciencia de marginalidad

LEOPOLDO ZEA, Filosofía de la historia americana

Abstract This introductory chapter is devoted to presenting the conceptual framework of economic vulnerability and its drivers, as well as analyzes the role of natural disasters in reducing assets accumulation and the derived regional development implications. It provides the background to understand the dynamic of economic vulnerability first in general and further concretely in Mexico. This chapter quotes a number of research works relevant to provide elements for a solid discussion on the original endowment in defining current social asymmetries, poverty and agricultural productivity in Mexico, coming up with an objective view of the problematic, and a grounded argumentation of this work's hypothesis.

Keywords Vulnerability • Resilience building • Coping capacity • Adaptive capacity • Hazard, stressor • Poverty • Rural development • Rural poverty • Migration • Exposure • Foreign trade • Assets • Entitlements

The chapter starts by providing a review of different concepts and approaches of vulnerability (Section 2.1) and economic vulnerability (Section 2.2) towards defining this work's concept of economic vulnerability. Further, we approach and briefly illustrate the structure of economic vulnerability and its functioning as a dynamic and cyclic process. One interpretation of the complex process of multiple interacting drivers of economic vulnerability is expounded in Section 2.3, emphasizing the role of assets, especially those of the poor in rural areas. This section also discusses poverty definitions, assesses briefly poverty in Mexico, and provides some elements towards explaining historic facts defining entitlements and some issues of economic policy behind poverty in Mexico and concretely in the countryside. The need for taking into account concepts and patterns of agricultural growth and rural-urban migration in a strategy to deal with economic vulnerability is presented in Section 2.4. External shocks tend to become amplified due to structural weakness and interacting stressors that are explicitly described within this chapter and summed

© The Author 2015
S.O. Saldaña-Zorrilla, *Natural Disasters, Foreign Trade and Agriculture in Mexico*,
SpringerBriefs in Environmental Science, DOI 10.1007/978-3-319-17359-7_2

up further below. We then, in the forthcoming chapters of this research work, leave the mere conceptual discussion and proceed to assessment of vulnerability and public-private response in Mexico, along with its possible solution paths.

2.1   Vulnerability

Vulnerability is the key concept to our approach, and we proceed now to expound definitions of vulnerability from different disciplines and perspectives relevant to the present work, their linkages to related concepts and, based on them, to propose a definition of economic vulnerability.

In a broad sense, vulnerability is incumbent upon two sides: unit of exposure and external force(s). So, vulnerability can be initially defined as the susceptibility of a certain unit to a specific force, and risk can be expressed as the probability of an undesired derived outcome, based on the potential occurrence of harmful events and on the susceptibility to them among those likely to be exposed (Dilley and Boudreau 2001).

The way sustainability science analyzes both elements included in these sides, as well as the complexity of their interrelations, provides an excellent basis towards defining, further, economic vulnerability. So, for some scholars of sustainability science, vulnerability is conceived, above all, as a coupled human-environment systems interaction, which possesses a likelihood to experience harm due to exposure to a hazard (Turner et al 2003).[1]

So, vulnerability does not involve merely active and passive factors, but rather dynamic objects and subjects in continuous motion. As pointed out in Turner et al (2003), systems have different sensitivities to perturbations and stressors strongly linked to entitlements in the case of social units.[2]

Along with entitlements, other elements defining susceptibility of social units are *coping* and *adaptive capacity*. Coping capacity can be defined as the ability of a unit to respond to a harm occurrence as well as to avoid its potential affectation. Adaptive capacity is the ability of a unit to gradually transform its structure, functioning or organization to survive under hazards threatening its existence (Kelly and Adger 2000). Another concept contributing to vulnerability comprehension is *resilience*, which, borrowed from physics, defines a system's ability to return to a reference state after a disturbance and to maintain basic structures and functions despite disturbance (Norris et al 2008; Turner et al. 2003). Thus, the expounded linked

---

[1] In this concept, hazards are understood as threats to the system, which act by means of perturbations and stressors. Perturbation is a major alteration in the system -of external origin- generating exceeding effects to those the system can cope with, and stress is a continuous increasing pressure upon the system. The novelty of this concept is that it expands the analysis spectrum to embrace multiple stressors and the structure of a hazard's causal sequence as a complex of socioeconomic conditions and biophysical subsystems lying behind.

[2] *Entitlements* are essentially the system of legal and customary rights defining access to society's resources.

concepts obligate us to incorporate social, economic, institutional, and cultural structures into the set of forces shaping a units' susceptibility in the vulnerability analysis, thus overcoming the eventual limitations of other frameworks, i.e. risk-hazard and pressure-and-release models.[3] Moreover, these concepts will be useful to the present work, since their implementation into our analysis may allow us to maneuver with existing structures in the country with which our case of study deals, especially in the part concerning decision making.

Vulnerability analysis is increasingly being applied to different concrete matters using implicitly most of the above listed elements of vulnerability, but emphasizing the concrete area they deal with. For instance, some works focused on human health vulnerability emphasize the lack of post-floods management in dengue outbreaks (Zapata-Martí and Saldaña-Zorrilla 2009); Works focused on concrete biophysical dimensions of vulnerability, i.e. water, emphasize, for instance, inadequate supplies of potable water and sanitation, and exposure to waterborne diseases of a given human settlement (Cocklin 2002); Works on vulnerability to climate change under-line the contribution of natural disasters mismanagement to increase vulnerability of human systems to natural hazards (Conde and Saldaña-Zorrilla 2008; Abramowitz et al. 2002); Other works on food security emphasize shortages in food supply due to extreme events (Liverman 1990). Approaches such as dynamic vulnerability focus on, i.e., the extent to which environmental and economic changes modify the capacity of regions, sectors, ecosystems and social groups to respond to shocks (Leichenko and O'Brien 2001).

2.2   Economic vulnerability

The rise of literature on vulnerability issues over the last two decades has demanded a more specialized definition of vulnerability. For instance, *social vulnerability* con-ceives social entities as its units of analysis, aiming concretely at identifying the way human societies and individuals are prone to suffer from disturbances and stresses (Adger 1999). So, social vulnerability, as expounded above, relies to a large extent on concepts like endowments and entitlements.

For its part, economic vulnerability is still sparsely defined, and one can find concepts which, though useful and valid, fail in describing vulnerability in its wide spectrum. There exist interesting views about economic vulnerability highly biased to the side of drivers, like those provided by some scholars of political economics in the context of international development. The germinal works on this matters of

---

[3] Risk-hazard models tend to consider impacts of hazards as a function of exposure to the hazard event and the sensitivity of the unit without clarifying how the units amplify or attenuate the impacts, as well as the role of multiple stressors in defining susceptibility. In the case of pressure-and-release models (PAR), they emphasize the conditionings of a given unit's lack of safety, including even ethnicity, class, etc. However, the PAR model does not incorporate biophysical subsystems interacting with society.

Todaro (1982) consider vulnerability as a situation in which least developed countries (LDC) find themselves in a dominance and dependence relationship *vis-à-vis* the developed countries. In this concrete view, LDCs are said to be economically vulnerable to the decisions of rich nations in areas such as trade, private foreign investments, foreign aid, technological research and development, etc.[4]

In line with Todaro's definition, the United Nations Conference on Trade and Development (UNCTAD) defines economic vulnerability as the structurally more exposed position of LDCs than most other developing countries to external economic shocks. Also, UNCTAD points out that economic vulnerability implies consequences of major global and regional economic and financial disturbances and increases in the prices of critical imports such as energy products; The typical export dominance of a single commodity or service sector makes their economies particularly vulnerable to adverse physical or economic shocks (UNCTAD 2003).

So, in the early 1990s, UNCTAD developed a first attempt to construct an index of economic vulnerability[5], and in 1994 the Programme of Action for the Sustainable Development of Small Islands Developing States (SIDS) adopted an index of economic vulnerability, expected to demonstrate that SIDS were generally more vulnerable than other developing countries.[6]

In the view of Briguglio (2002), a country can be economically vulnerable and yet register a relatively high GDP per capita. So, countries like the SIDS are particularly economically vulnerable due to their limited ability to exploit economies of scale, lack of natural resources, low diversified economy, dependence on narrow range of exports, and high dependence on imports of strategic goods, i.e. fuel and food. Notwithstanding, what essentially makes a country economically vulnerable in the definition of Briguglio, is its exposure to economic forces outside its control. Thus, the *peripherality* condition of an economy goes beyond geographic insularity and remoteness (leading to high costs and marginalization from world trade), but also includes inability to influence international prices (price-taker economies).

However, being vulnerable is not only a question of poverty and smallness of a country, as this work approaches. Vulnerability accrues also to countries of big population and large economies, whose vulnerabilities are less visible at a glance, and only through more detailed analysis exhibit differential vulnerabilities due to dualistic characteristics (Rodríguez 1980). So, above all, Latin American countries

---

[4]This is a useful concept whose asymmetry component is closely interconnected with other concepts from the economics of development, like the center-periphery relations and terms of trade in the works of Raúl Prebisch (i.e. 1950 and 1973, respectively).

[5]*Cfr.* Briguglio, L. (1992). Preliminary study on the construction of an Index for ranking countries according to their economic vulnerability. Report to UNCTAD, 1992.

[6]The UNCTAD Economic Vulnerability Index was constructed as a composite indicator based on three fundamental dimensions: (1) the magnitude of external shocks beyond domestic control (measured through indicators of the instability of agricultural production and exports); (2) the exposure of the economy to these shocks (estimated through the share of manufacturing and modern services in the gross domestic product, and an indicator of merchandise export concentration), and; (3) the structural handicaps explaining the high exposure of the economy (taking into account economy's smallness, measured by a proxy demographic variable) -UNCTAD 2003.

like Mexico, Brazil, and Argentina should not be considered as entirely vulnerable, but unequally vulnerable, whose rich and poor societies, high productive and left-behind economic sectors, etc. coexist at differential degrees of vulnerability (Rodríguez 1980; Colosio-Murrieta 1979).

More recently, economic vulnerability is being used to refer to the extent to which macroeconomic policies can exhibit performance inconsistencies, sudden loss of net national product, and hence lead to economic crises, as a result of underscored development of warning systems and an economy's inability to work with multiple equilibria (Yap 2002). The response to these types of vulnerabilities can consist of financial monitoring and modeling of early warning systems, as Kaminsky and Reinhart (1996) propose to concretely avoid economic crises originated in financial factors like the Asian one of 1997. However, these approaches do not address structural factors of the economy, and tend to reflect rather what this work calls risk and preparedness, instead of vulnerability and vulnerability reduction, respectively.

2.2.1   Coping and adaptive capacity

In defining economic vulnerability it is crucial to consider approaches with imbedded structural factors. So, the implications of vulnerability to economic analysis are explicitly expounded by Amartya Sen (1981) by relating entitlements and initial endowment in a coherent process where individual levels of vulnerability are conditioned by broader institutional structures, which sometimes reproduce or even amplify vulnerability given society's capacity to provide opportunities. These views and concepts are inserted below both in our immediate concept of economic vulnerability as well as in further details in Section 2.3. Now, let us explain the concrete elements constituting coping and adaptive capacity in economic vulnerability.

I. Coping capacity

The capacity to respond to a harmful event as well as to avoid its potential affectation of an economic agent is not only determined by its productive level, but also by its relative position within its society. Factors like entitlements, information availability and assets distribution in a society gives shape to this coping capacity.

  i) *Entitlements.* Connected with this way of seeing vulnerability from the perspective of economic agents, Amartya Sen (1981) addresses vulnerability using entitlements. Entitlements are the package of goods and services, which an economic agent can obtain by means of trading their stakes under current regulatory conditions. In other words, there exists an intrinsic susceptibility in the way economic agents profit from their assets.[7] In addition, entitlements also influence information availability, which plays a key role concerning economic vulnerability, since the common assumption of homogeneity of information

---

[7] With a very similar meaning, Cannon (1994: 19) calls this *vulnerability of livelihood resilience.*

among economic agents (as in the neoclassic approach of economics) does not usually apply in practice. Heterogeneity with respect to information possession makes considerable differences when taking decisions and in turn when obtaining incomes. Better informed economic agents are normally more able to identify the risk their asset implies and hence to take better decisions.

ii) *Assets*. Assets are a key variable to understand impoverishment in poor rural families (or households), which can be defined as the stock of wealth used to generate well being (Vatsa and Krimgold 2000). This concept is important when considering the effects of natural disasters, which can decrease the capital assets of households and businesses. Families have an initial asset, which generates an output. This output varies widely, depending on market price of the produced factor, and on the productivity of its use (profitability). As families pursue strategies to maximize their assets, they are in better position to enlarge their risk pool and reduce vulnerability.

II. Adaptive capacity

The ability of a unit to gradually transform its structure and/or functioning to survive hazards affectation is interconnected with assets and entitlements as well in that it influences economic agents' ability to hedge from adverse events. Explicitly, adaptive capacity concerning economic vulnerability includes risk management and protection actions.

i) *Risk management*. Derived from options to deal with negative shocks, prevailing risk management capacity in a country is crucial in reducing vulnerability. Risk management is the set of pre- and post-disasters actions towards facing negative consequences of hazards (Freeman and Mechler 2001). Pre-disaster actions include risk transfer (i.e. insurance), risk assessment (monitoring, mapping, etc.), mitigation works (i.e. infrastructure strengthening, etc.), and preparedness. Post-disaster actions include emergency response (aid, clean-up, etc.), rehabilitation, rebuilding, and loss sharing (i.e. private-public, national-local).

ii) *Protection*. On its part, protection is a concept situated at both sides of risk management phases (pre- and post-disaster). Protection is the network of awareness and emergency response for disaster and crisis state avoidance, initially based on various technical interventions usually known as preparedness.[8]

In concordance with the conceptual discussion above expounded and attempting to overcome dispersion and partiality of available definitions of economic vulnerability, this work defines economic vulnerability as the susceptibility of an economic agent to absorb external shocks (hazards) negatively, given its assets possession and

---

[8] Preparedness is the management capability before a disaster occurs to provide an effective and efficient (prompt and due) reaction to face a disaster (Freeman et al 2002). Protection granted by governmental planning plus those from other social institutions are termed social protection, which acts complementary to self-protection (Cannon 1994).

entitlements system (coping capacity), as well as its implemented risk management and protection measures (adaptive capacity). Finally, reducing economic vulnerability consists basically of implementing the due changes in time to minimize negative effects from exogenous shocks upon economic agents' assets by strengthening contractual elements in the society (entitlements and assets) and/or improving the risk management and protection strategies.

2.2.2 The cycle of economic vulnerability

Figure 2.1 below provides a very simple illustration of, first, the structure of economic vulnerability according to this work's approach, as well as of some elemental cause-effect relations derived from hazard occurrence. An individual unit of analysis' vulnerability is depicted in the diagram by three fundamental elements: Unit adaptability, public response, and probability of economic losses. Unit adaptability embraces coping and adapting capacity within the framework of society's prevailing entitlements and assets distribution.[9] The second pillar of economic vulnerability is public protection, understood as the set of social programs, antipoverty strategy, etc., constituting a social and redistributive policy, as well as the public mechanism of disasters avoidance, whose strength or weakness degree contributes to define economic vulnerability of both our unit of analysis as well as to the entire society –elements of risk management are marked with an asterisk in the diagram.

The share of assets at risk –out of units' assets total value- defines the probability of economic losses. Assets at risk are the monetary valuable elements, i.e. capital stock, production of goods and services, and in general any form of assessable wealth, exposed to hazards to a given extent. Economic Vulnerability is the outcome of these factors together. To this extent of the diagram, all said elements are potential facts, which materialize until a hazard hits the system.

The hazard causes damages that are translatable into economic monetary losses, whose implications can be direct damages, like loss of public and private productive infrastructure, housing, etc. Also, such a loss of assets leads to indirect losses,[10] to productive disruption and, therefore, to decreasing incomes of private economic agents. The government, for its part, may decrease revenue from state-owned companies as well as because of decreased taxing basis (concerning basically income and consumption taxes) as a consequence of a potential reduction in economic activity. This process produces a vicious cycle as less income leads to less financial

---

[9] In fact, adaptability is the preferred term for this work though that word itself could seem to misrepresent its components, and perhaps could be better represented by terms like *resistance*, *strength*, etc. However, most literature of vulnerability uses it conventionally to represent system's ability, competency or capacity of a system to adapt and respond to climatic stimuli. Also, see Schjolden (2003), Kelly and Adger (2000), and Chambers (1989).

[10] Indirect losses are actually the share of unrealized production, which act by reducing GDP growth in the current year, whereas direct losses are not accounted into GDP growth reductions, since they are added value belonging to GDP accounting of past years (cfr. ECLAC 2002).

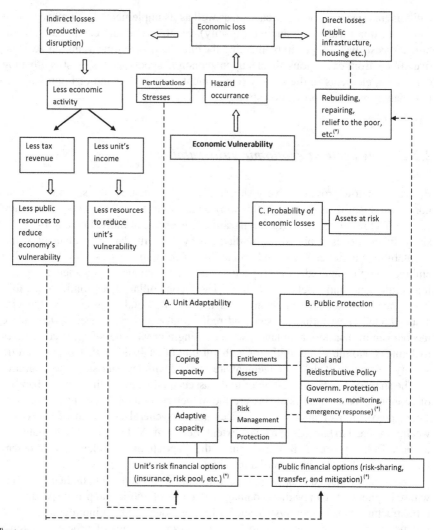

Fig. 2.1 The structure of economic vulnerability and its cycle

ability to invest in reducing vulnerability both at unit level, i.e. household, as well as at global level, i.e. economy, in reducing vulnerability. Unlike perturbations, stresses are linked to the unit's resistance (discontinuous line) because, as above expounded, they are understood as threats to the system, which act by means of continuous increasing pressure upon the system, and are associated to a large extent to structural conditions. Appendix A provides further elements to analyze the dynamic of economic vulnerability, suggesting a mechanism to maneuver with coping and adaptive strategies in order to reduce economic vulnerability under this work's framework.

2.3   Drivers of economic vulnerability

So far, the basic elements shaping economic vulnerability have been presented here, and we turn now to explain the relevance of the hazards analyzed in the following chapters of this book (natural disasters and trade liberalization). Considering that hazards are threats to the system, which act by means of perturbations and stressors, analyzing assets structure and their functioning along with the historic impact of the entitlements in this country is crucial to clarify why natural hazards and economic policy have been drivers of increasing vulnerability in Mexico.

2.3.1   Assets accumulation

Increasingly, scholars argue that poverty is not only a lack of income or consumption, but also a lack of assets (Haveman and Wolff 2000; Oliver and Shapiro 1990; Sherraden 1991). So, asset poor involve those households with insufficient resources to invest in their future or to sustain household members at a basic level during an economic disruption (Fisher and Weber 2004). There exists empirical evidence showing that wealth is more unequally distributed than income. For instance, Wolff (2001) points out that the top 20 percent of households in the USA earn 56 per cent of the nation's income and own 83 percent of national wealth. For that reason it is crucial to identify poor households' assets in order to, further, encourage assets accumulation.

Among other authors, Chambers (1989) cautions about the relevance of increasing assets in low-income families, since this improves human conditions beyond poverty just in terms of flows, but also structural vulnerability. He affirms that vulnerability is even more interlinked with net assets than poverty. For authors like Vatsa and Krimgold (2000), vulnerability is a broader and more dynamic concept, which involves the poor, but also households living above the poverty line at risk of falling below in case of an income shock (new poor). Given that linkage, factors that obstruct an accumulation of assets are, in turn, impeding poverty reduction and putting additional population into poverty. For instance, losses from natural disasters or income reductions due to depressed agricultural prices impede rural households in accumulating assets, creating a vicious cycle of inefficient risk management strategy, low return, low consumption and low savings and investment (Vatsa and Krimgold 2000).

A poverty reduction strategy possesses two sides. One consisting of irreducible actions aimed at providing short-term relief to families living in poverty (i.e. health facilities) and making affordable some long-term intangible assets (i.e. education). The other side is related rather to directly strengthen income generation, i.e. agricultural commercialization, reducing information asymmetries, productivity enforcement, etc. However, the latter side of the anti-poverty strategy in Mexico –as in most of the countries- is less actively promoted (Fisher and Weber 2004; ECLAC 2001;

Attanasio and Szekely 1999). It is so partly because it implies actions falling outside the jurisdiction of mere poverty-related governmental agencies. It is so also because strengthening income generation is more complex and usually requires changes to the national socioeconomic structure and lot of political lobbying and willingness.

Programs aimed at reducing poverty tend to increase transfers, but they do not affect the long-term lack of assets in the current case of Mexico (Attanasio and Szekely 1999). Anti-poverty programs help to raise income or consumption but only in terms of flows. In addition, ECLAC (2001, p. 32) points out that social programs of poverty reduction in Mexico are merely of aid character, though if these programs were more substantial and long-term sustained, these additional resources could increase assets, as well as for instance improve ability to invest in education and other long term variables, which directly modify assets and in the last instance income. For that reason, prevention/compensation mechanisms are required. In Mexico, such mechanisms have been actively promoted by the State during the past 90 years, as part of the resulting social State implemented by post-revolutionary governments.[11]

2.3.2   Poverty and entitlements

Stressors, understood as continuous increasing pressure upon a unit, are analyzed in this section when approaching entitlements determinants in prevailing poverty in Mexico, i.e. the colonial legacy and industrialization patterns during the XX century. The following subsections are an attempt to define poverty for this work's sake based on the vast conceptions and measurements of poverty in order to further present the most likely causes of structural poverty in this country.

Defining poverty implies dealing with a very controversial paradigm. There exists no consensus about what issues are embraced by poverty. However, a point of agreement among the vast existing literature on poverty seems to be that poverty is a lack of opportunities to reach a basic minimum level of well-being. In turn, well-being can also include several determinants, i.e. assets, infrastructure, social networks, institutions, human development, etc. This leads us to another point of agreement: poverty is a relative concept that deals with "minimums of well-being" (as in SEDESOL

---

[11] A key argument in the current debate about development in developing countries is that income alone is an insufficient indicator of economic well-being. In response, various alternative supplementary measures have been proposed, including consumption-, income- and wealth-based indicators (SEDESOL 2002; Haveman and Wolff 2000; Slesnick 1993). The present work considers assets the ideal welfare measure. Despite recognizing that fact, this work approaches assets by household incomes in measuring economic vulnerability in chapter five. We did so because available statistics and data sources do not allow us to make assets' quantitative analysis. Hence, quantitative analysis of economic vulnerability until assets level will be possible for only the individual municipalities that our case of study inquires through field work in chapter 4.

2002), "society's hopes and aspirations to be fulfilled" (World 2001a), or "basic needs" (Todaro 2000), all varying widely among and within societies.[12]

Other concepts of poverty go beyond tangible factors and include rights issues, relationships, powerlessness, socioeconomic exclusion, and even loss of dignity (i.e. Elankumaran et al. 2000). Nevertheless, this work will only focus on those more quantitatively measurable economic dimensions of poverty.[13]

The World Bank 2004 Report on Poverty in Mexico estimated income poverty in Mexico at 51.7%, and consumption poverty at 51.3% of the population in year 2002 (World Bank 2004). In addition, the country has clear gaps in terms of purchasing power. According to the Poverty and Social Development Indicators of the Mexico's Country Assistance Strategy (World Bank 2002), inequity in power of consumption is dramatic: Consumer price index shows 219, compared to food price index 227 (1995 = 100). Consumption in terms of income distribution reveals also high inequity in Mexico: the lowest (poorest) deciles consume only 4% of income, whereas the highest one consumes 56.7% (INEGI 2000; World Bank 2001a, 2001b).

However, both moderate and extreme poverty in Mexico is dramatic in rural areas. 74% of rural population lives in poverty -in urban areas it represents 36% (World Bank 2002). For its part, 20% of the Mexican population is reckoned to live in extreme poverty,[14] from which 65% are in rural areas (World Bank 2004: 54). Also, the livelihood of 72% of the extreme poor in Mexico is agriculture (World Bank 2004: 54). Section 2.4 below presents a more detailed description of rural livelihoods and their links to vulnerability.

Low education is, in turn, much more concentrated in rural areas: 73% of rural extreme poverty has no education or incomplete primary –compared to 51.3% in urban extreme poverty. Nevertheless, it is still risky trying to explain income differences by merely education asymmetries. For instance, based on a regression analysis, the World Bank staff in Mexico estimated the contribution of household characteristics to rural-urban income differences, and reckons that 35% of that difference is due to disparities in educational levels, but employment characteristics explain ca. 50% of that difference (World Bank 2004: 57). That analysis compared employment characteristics between remunerations from agricultural and

---

[12] For instance, infrastructure necessities in least developed countries consist conventionally on tap water and sanitation services, but in middle income countries like Mexico electricity is included as well.

[13] Among measurements of poverty, there are two basic criteria: income and consumption. *Income-based* measurements of poverty embrace the total current monetary and non-monetary inflow a household obtains from different functional sources, including wages, salaries, dividends, rents, etc. The use of income-based measurements is advantageous in that of allowing identifying those economic activities to be strengthened to reduce vulnerability, -tough more subject to errors due to difficulties in incorporating non-monetary income, self-consumption and transfers. *Consumption-based* measurements consist of the total amount of monetary and non-monetary expenditure a household makes in all possible items. The advantage of using consumption-based measurements of poverty relies on the fact that it reflects smoothing strategies a household implement in response to sudden income fluctuations, i.e. family solidarity, aid, remittances (World Bank 2004).

[14] Extreme poor consists of populations living below the food-based poverty line.

non-agricultural activities, attempting to reflect differences in a household's living standard of being employed in industrial sectors relative to the agricultural ones.

Inequity in rural areas has increased throughout the last decade. Based on the Gini coefficient,[15] the rural expenditure coefficient increased from 0.41 to 0.48 between 1992 and 2002, reaching its top in year 2000 with a 0.56 coefficient (World Bank 2004). Also, despite the fact that over 25% of the Mexican labor force works in agriculture, only 2.5% of total insured people by the Mexican Social Healthcare Institute (IMSS) work in this economic sector (INEGI 2003).

As in other Latin American countries, poverty in Mexico is a complex issue with ancient causes (e.g., land tenure) but also explained by dynamic factors and processes (e.g., economic policy and extern shocks). The former corresponds to the so-called original endowment[16] in the form of initial allocation of wealth -after the Spanish conquest in our case of study. Factors like the industrialization model undertaken during the 50's and 60's provides some elements to understand poverty increase in the countryside.

On the one side, initial endowment during the colonial establishment margined indigenous population of wealth property, making them serve as slaves (*Encomienda System*) –even though some indigenous communities could work its arable land under permission and contributing to the colonial authority's revenue. It worked in that way until the independence war (year 1810), when slavery was abolished, but even in spite of that fact, the most productive lands were already appropriated and concentrated in very few hands. So, at this first stage of independent life, dispossessed but now free farmers had to continue being still highly exploited by semifeudal productive systems. It was so basically because the resulting government of the independence war did not change the *status quo* of former wealth allocation derived from a sort of amnesty pact between insurgency, Catholic Church and colonial officials as pacification and independence conditions (Alaman 1968). The Mexican Revolution forced land redistribution (Aguilar and Meyer 1989), and a successful and gradual process was undertaken by the early post-revolutionary governments[17] to improve conditions in the rural poor (Cosío-Villegas et al. 1991) and to eliminate the *Hacienda* system. But from the end of the 1940's, industrial policy in Mexico seems to have been reoriented in favor of the industrial sector (Tejo 2000: 10).

The school of structural economics supported reorientation from an agricultural to an industrial driven economy in Latin-America. Raúl Prebisch (1951), one of the initiators of economic structuralism, points out that if one considers the high share

---

[15] The Gini coefficient is a measure of income inequality developed by the Italian statistician Corrado Gini. The Gini coefficient is a number between 0 and 1, where 0 means perfect equality (everyone has the same income) and 1 means perfect inequality (one person has all the income, everyone else has nothing).

[16] For further details about Endowment, and Endowment Effect, see John List (University of Maryland), and Daniel Kahneman (*Homo sapiens vs. Homo economicus*).

[17] Post-revolution is conventionally considered the period between the end of the government of Alvaro Obregon –the last *Caudillo*-, in 1928, and the end of the Second World War in 1945, characterized land redistribution, expropriation of the oil industry and the beginning of a model of economic growth based on imports substitution.

of working population employed in agriculture in Latin America, one can understand that land tenure is only part of the big issue about economic development. For him, any solution must first take into account that there will not be any improvement in people's standard of life, as long as exceeding workforce in agriculture is not eliminated and redirected to higher productivity sectors. According to that efficiency approach, the sector expected to absorb such a displaced labor force was to be Industry.[18] So, the economic pivot would be industry, which would become the most dynamic sector, able even to encourage agriculture by demanding increasingly raw materials (Figueroa 1991). Under these assumptions, agriculture would turn into an input provider for Industry. Proof of that are policies to deflate agricultural prices –especially those of primary goods- in order to encourage industry by means of making affordable industrial inputs. In practice, inability to sustain such an industrial policy in the long-term in Mexico led to failing in turning theory into reality (Montserrat and Chavez 2003).

During the 1950s, the first scholars of modern economic development identified industrialization and urbanization as the main structural change a country should undergo in order to improve welfare and for economic growth (Lewis 1954; Fein and Ranis 1961). The core of that argument is that, historically, economic systems are composed of sectors characterized by a clear difference in factor endowments. In such case, it is possible to shift factors from less to more productive sectors. Historically, this shift has taken place from agricultural to non-agricultural activities, and labor is the most frequent factor in motion.[19] This reallocation of factors leads also to a rise in efficiency of food production, creating thus an agricultural surplus, which in turn provides the basis for fast industrialization and, further, for growth and poverty reduction. In addition, the tendency of higher relative value of manufactured to agricultural goods, known as the *Prebisch-Singer Theory of Terms of Trade* (Prebisch 1950, 1951; Singer 1950), is another crucial reasoning to base economic growth on Industry.

Contrary to theoretical expectations, labor transfers from rural to urban areas in Mexico have exceeded economic systems capacity of employment. The cost of urbanization is high when urban conditions are not appropriated to absorb additional workforce. Population in Mexico became prematurely urbanized in the sense that the share of urban population was greater than the current stage of development could support (Colosio-Murrieta 1979). Global crisis during the 70's, oil crisis, high cost of public debt (due to the rise in international interest rates), along with wrong estimations on future country's incomes, slowed down industrial production in the 1980's, accompanied with an increasing carelessness about agriculture. As a result, the industry did not develop as expected and therefore was unable to absorb an exceeding labor force, which in fact had been released from agricultural activities.

---

[18] Industry understood as the group of activities related to primary resources processing and manufactures.

[19] For instance, the historic process of recognizing property titles to landlords in Britain embracing communal lands during the XVII and XVIII centuries forced displacements of enormous masses of rural workers to urban centers to meet the increasing workforce demand from the increasing industrialization process (Marx 1867; Ricardo 1817).

Below in chapter 3, we expound on the state of the art of agriculture in Mexico, the process of trade liberalization gradually implemented from, roughly, 1986 and abruptly accelerated in 1994 when NAFTA started, as well as discuss the impact of trade liberalization on incomes and poverty in the countryside.

2.3.3   Emigration as response to rural poverty

Lack of support for the countryside and agriculture, as well as inappropriate urban projecting in large cities in Mexico, has contributed to increasing urban vulnerability after the 50's. There are theories of rural-urban migration helpful to understand better the phenomenon of slum concentrations in urban areas (Lewis 1954; Fein and Ranis 1964; Todaro 2000), most of which find the labor market failure in the (1) incapacity of agriculture to reabsorb rural workers together with its respective low-income (productivity), and (2) insufficient industrial development in the cities. For Todaro (2000: 305), migration is primarily an economic phenomenon, which for the individual migrant can be a quite rational decision despite the existence of urban unemployment.

The Todaro model postulates that migration proceeds in response to urban-rural differences in expected incomes rather than actual earnings. That decision is taken in order to maximize their expected gains in longevity and, for a given time horizon the urban sector results are more convenient. In that sense, as natural disasters reduce future incomes expectations from agricultural activities, it stimulates in turn slums growing as well. Empirical evidence in that way was found by this research work, and is presented in chapter 3 when discussing vulnerability in Mexico City, in chapter 5 when making spatial econometric analysis at municipality level, as well as verified in the stakeholders' consultation in three communities in rural Mexico, presented in chapter 6 of this work.

In last instance, such failures in rural-urban incomes forecast have drawn a vicious circle of public inefficiency. As migrants from rural areas do not get employed in urban industrial activities, they are forced to engage in low-productivity tertiary activities. Also, large a portion of this underemployed population becomes an obstacle to an efficient allocation of public resources, since society is forced to provide large amounts of urban social infrastructure at the expense of direct public investments. For that reason, investments in agricultural productivity as well as in improving rural income can provide valuable solutions both in fighting poverty and in allocating efficiently society's resources.

2.4   Agriculture as livelihood

Despite producing about 4% of GDP in 2002, the agricultural sector employs over 20% of the national population and its relative economic weight has declined permanently throughout the last five decades –in 1950 it provided about 18% of GDP.

Agriculture is still the main livelihood of the poor in rural areas in Mexico (ECLAC 2001). However, agricultural income is increasingly complemented by income from low-salary activities. In general terms, poor rural population work in small-size rainfed agriculture, producing basic grains of fluctuating low prices, with inability –rather than aversion- to contract crop insurance.

The typical landholding in the Mexican agrarian system is the *Minifundio*, which is too small to provide the workers with levels of living much above the bare survival minimum. For that reason, holders of *minifundios* are forced to provide seasonal labor to *latifundios*, or even to other labor markets, working either as wage laborers in other agriculture activities nearby, or taking a job in non-agricultural activities in semi-rural areas. It happens regularly because agricultural incomes are so low, that it is not enough even to maintain subsistence. That fact was empirically verified by this book's field work, and is presented when analyzing subsistence farmers' coping strategies in Southern Mexico. In sum, only 45% of total income of rural workers in Mexico is provided by agriculture, with a decreasing trend throughout the past 30 years (ECLAC 2001).

As land-tenure is smaller, a higher share of personal income must be obtained from alternative sources, as shown in an empirical study of the Mexican case carried out by Janvry and Sadoulet (2000). Owners of two hectares or smaller properties earned only 23% of their income from agriculture in year 1997, whereas holders owning more than 18 hectares earned at least 62% from this activity. Also the dependence from remittances and self-employment decreases as the size of land increases.

However, shifting economic activity may fall outside the coping capacity of a rural household, given that it uses to lead to destruction of its livelihood, in many cases forcing a change to the unit's livelihood permanently. The latter is relevant to the present work given that it makes visible that as agricultural income decreases, the coping capacity of the worst-off farmers comes to an end, and they have then either to share income source or even to shift activity. So, protection of rural households must not be thought as facilitating the transition from rural to urban economy, but rather strengthening rural incomes. For that reason, the present work includes field week with Mexicans farmers living in extreme poverty, and attempts at getting a realistic picture of their circumstances as possible, as well as to inquire into their coping strategies, in order to incorporate that learning to come up with stakeholders-based solutions.

Increasingly, rural incomes rely to a larger extent on transfers (from social programs, remittances, etc.) as part of income. At national level, remittances from nationals working abroad represented 1.2% of households' total income in year 2002. However, remarkable differences arise when comparing among urban and rural, where it reaches 0.5% and 3.5%, respectively. Particularly relevant are remittances to the first rural quintile (poorest 20%), where remittances represent 20% of the income. That quintile is reckoned to obtain 53% of total income from remittances, social programs, and other transfers (World Bank 2004:75).

Given current inequity conditions in Mexico, agricultural growth itself is not enough to reduce poverty in Mexico. In projecting towards reducing both economic vulnerability and poverty in the countryside, it is crucial to account for some structural patterns of agricultural growth, whose characteristics may help to maximize benefits from public intervention. Agricultural growth does not necessarily lead to poverty reduction in all countries. Empirical studies show remarkable progress in poverty reduction after agricultural growth under certain particular conditions: It depends on income structure and other distributional variables in a given country. The Ravillon and Datt (1996) results about India show that 85% of poverty reduction in the country for the analyzed period is explained by agricultural growth. Also, rural absolute poverty is negatively related to rural real incomes and the average crop yield, which is also an increasing function of food prices.

For Timmer (1997), agricultural growth and poverty reduction depend on equity state. If national growth were uniform (elasticity of connection = 1) in all income-ranges, then simple growth would generate poverty reduction at the national growth rate. But in practice, countries' structure is not uniform. On the one hand, some countries have an elasticity-growth greater than 1 in high-income population, whereas less than 1 in the case of poverty, like agriculture in Pakistan (Adams and He 1995). On the other hand, countries with the inverse behavior would benefit even from neutral support policies to agriculture. In this sense, agricultural growth contributes to both growth and poverty reduction if (as discussed in Sarris 2001) the achievements of the initial productivity stimulus is allocated on economic agents who, in turn, circulate this flow by investing/spending domestically on labor intensive domestic products. Timmer's analysis (1997), based on rural and urban income by productive sector, shows that for countries with large income gaps, growth in agricultural activities is no more successful in poverty alleviation than growth in the non-agricultural economy. Mexico is situated in this case. Under that analysis, one must not be expected to achieve large progress in poverty reduction in Mexico by means of only improving average productivity in the countryside, but exclusively by aimed strategies. In that sense, one must admit that even when the current strategy to reduce natural disasters risk in the countryside and to support agricultural production provide valuable support, it is not optimally contributing to reduce poverty given the current structure of income in the Mexican countryside. To meet with that, support to reduce vulnerability to natural disasters in the countryside in Mexico must discriminate against different economic agents. Public support (and subsidy) to increase productivity in the countryside must keep in proportion with a necessity degree. For instance, Chile has achieved a more equitable system to provide subsidies for crop insurance premiums, by supporting up to 80% of premium to those poorest farmers, whereas the support reaches at maximum 50% to farmers with more financial ability and better market information. By contrast, subsidies to crop insurance in Mexico have benefited relatively more to non-poor farmers, as criticized in chapter 4 when this work turns to assess the current disaster management strategy and concretely of *AGROASEMEX*, *Fondos* and *FOPREDEN*.

2.5   Final comments

Analysis of economic vulnerability implies dealing with a complexity of elements interacting on a given unit to deal with hazards, demanding a careful identification of what makes it vulnerable. A hazard, generally expressed as threats to a given system, affects depending on strengths and weaknesses of the unit, defining its coping and adaptive capacity. In the economic ambit, hazards can embrace abrupt perturbations derived from economic crisis, economic policy reorientation, among others, or can arise from relatively subtle stressors, like unequal entitlements and assets distribution, which puts continuous pressure upon the unit. In the natural ambit, hazards can also be subtle as well, as the case of continuous reduction in rainfalls, recurrent heavy winds and rains, etc., or more abrupt like their corresponding superlatives, i.e. droughts, hurricanes, etc.

Coping capacity in Mexico, understood as a unit's ability to respond to a given hazard to reduce its affectation, is being highly undermined by unequal access to society's resources and prevailing poverty conditions, more remarkably in rural areas. The impossibility of rural masses to accumulate assets seems to be contributing to reduce their capacity to cope with negative outcomes from natural disasters and current economic challenges. In addition, social programs in Mexico, though valuable, do not show clear signs of being contributing substantially to reduce economic vulnerability. In fact, even if the social policy works correctly in this country –tested further in chapter 5-, prevailing entitlements and assets distribution systems seem to reduce their efficacy. Historic unattended social and economic demands continue threatening incomes, especially those in rural areas, whose negative effects can be observed in the increasing emigration tendency of the last years.

As outlined in the discussion on economic policy throughout this chapter, one has to admit that a country with real development aspirations cannot rely on agriculture to grow its economy. However, if the urban economy fails to provide enough employment to additional workforce arriving from the impoverished countryside, given low dynamism of the national industry –as over the past 20 years in Mexico, an emerging policy to strengthen rural assets and manage disasters has to be more actively implemented. A country like Mexico, where a quarter of the national population relies on agricultural activities producing less than 4% of the national product, has to not only upgrade productive infrastructure to increase rural incomes, but also has to manage its disasters and market risk in a more efficient way. Even if hazards threaten especially the poorest population, one has to also not misunderstand and label the poor as vulnerable, since population above the poverty line is as vulnerable as the poor: its main risk consists of falling below the poverty line after an extreme hazard.

So far this work has justified the need for investigating natural and economic hazards, and the next chapter attempts to present more substantially natural disasters and trade liberalization affectation over the past two decades in this country in order to identify the stakeholders and involved institutions. This chapter's arguments concerning entitlements and assets and its relevance in shaping coping capacity will be complemented in Chap. 2, where adaptive capacity is addressed in more detail, assessing the scope of public intervention in social programs and risk management strategy. In general, chapters 2-4 are thought to gradually expand the spectrum of this vulnerability analysis to embrace multiple stressors and hazards' sequence with a complexity of socioeconomic conditions and biophysical subsystems behind.

Chapter References

Abramowitz J, Banuri T et al. (2002) Adapting to Climate Change: Natural Resource Management and Vulnerability Reduction. Background Paper to the Task Force on Climate Change, Adaptation and Vulnerable Communities. World Conservation Union – Worldwatch Institute - International Institute for Sustainable Development - Stockholm Environment Institute

Adams RH, He JJ (1995) Sources of Income Inequality and Poverty in Rural Pakistan. International Food Policy Research Institute. Research Report 102. Washington, DC

Adger NW (1999) Social Vulnerability to Climate Change and Extremes in Coastal Vietnam. World Development 27

Aguilar H, Meyer L (1989) A la sombra de la Revolución Mexicana. Mexico. Cal y Arena, Mexico City

Alaman L (1968). Historia de México. Mexico City, Ed. Jus, 1968, 5 Vol

Attanasio O, Szekely M (1999) La pobreza en América Latina. La pobreza basada en los activos. El trimestre económico, pobreza y activos en América Latina, Vol. LXVI (3), No. 236. Fondo de Cultura Económica (FCE). Mexico City

Briguglio L (2002) The Economic Vulnerability of Small Island Developing States. In: Sustainable Development for Island Societies: Taiwan and the World, Asia Pacific Research Program w/SARCS Secretariat Publication. Taiwan

Briguglio L (1992) Preliminary study on the construction of an Index for ranking countries according to their economic vulnerability. Report to UNCTAD 1992

Cannon T (1994) Vulnerability Analysis and the Explanation of 'Natural' Disasters. In: Varley A (ed) Disasters, development and environment. Willey, Chichester

Chambers R (1989) Vulnerability, Coping and Policy. IDS Bulletin 20:1–7

Cocklin C (2002) Cities, water and global environmental change: governance, water use and infrastructure as a remedy for water issues in cities. In: UPDATE IHDP, Newsletter of the International Human Dimensions Programme on Global Environmental Change. 03/2002

Colosio-Murrieta, LD (1979) Urbanization and Economic Development in Mexico. International Institute for Applied Systems Analysis (IIASA). Working Paper. Laxenburg, Austria

Conde C, Saldaña-Zorrilla SO (2008) Climate Change in Latin-America and Caribbean Countries: Impacts, Vulnerability and the need for Adaptation. Challenges on Climate Change in Latin-America. Foundation Futuro Latinoamericano (FFLA-Equator), The International Development Research Centre (IDRC-Canada) and Department for international Development (DFID-UK). 17 pp. January 2008

Cosío-Villegas D et al. (1991) Historia Mínima de México. El Colegio de México. Fondo de Cultura Económica (FCE). Mexico City

Dilley M, Boudreau TE (2001) Coming to terms with vulnerability: a critique of the food security definition. Food Policy 26

ECLAC (2002) Manual for Estimating the Socio-Economic Effects of Natural Disasters. United Nations Economic Commission for Latin-America and the. Caribbean, Mexico City

ECLAC (2001) Instituciones y pobreza rurales en México y Centroamérica (Institutions and rural poverty in Mexico and Central America). United Nations Economic Commission for Latin-America and the Caribbean, P. 30, Mexico City

Fein JC, Ranis G (1964) Development of the labor surplus economy: theory and policy.

Fein JC, Ranis G (1961) A theory of economic development. American Economic Review 51:533–565

Figueroa A (1991) Desarrollo agrícola en la América Latina (Agricultural development in Latin-America). In: El Desarrollo desde adentro. El Trimestre Económico No. 71. ECLAC. Fondo de Cultura Económica. Mexico City

Fisher M, Weber B (2004) Does economic vulnerability depend on place of residence? Asset poverty across the rural-urban continuum. Rural Poverty Research Center. Working Paper No. 04–01

Freeman P, Mechler R (2001) Public sector risk management in Mexico for Natural Disaster Losses. Issues paper for a Wharton-World Bank Conference on "Innovations in Managing Catastrophic Risks: how can they help the poor?" Washington, DC

Freeman P, Martin L, Bayer J et al (2002) National System for Comprehensive Disaster Management, Phase 2: Financing Reconstruction. Dialogue, Inter-American Development Bank Regional Policy

Haveman R, Wolff E (2000) Who are the asset poor? Levels, trends and composition, 1983-1998. Washington University Center for Sociial Devlopment. Working Paper 00-12. St. Louis

INEGI (2003) National Institute of Statistics. Geography and Informatics http://wwwinegigobmx/est/contenidos/espanol/tematicos/mediano/medasp?t=agro01&c=4790 Accessed 20(June 2003)

INEGI (2000) Anuario de Estadísticas por entidad federativa, Ed. 2000, National Institute of Statistics, Geography and Informatics, Aguascalientes http://www.inegi.gob.mx/difusion/espanol/bvinegi/aef/indef.pdf Accessed 15 December 2000

Janvry A, Sadoulet E (2000) Rural Poverty in Latin America: Causalities and Exit Paths. Food Policy 25(4):379–530

Kaminsky G, Reinhart CM (1996) The twin crises: the causes of banking and balance-of-payments problems." International Finance Discussion Paper No. 544 Washington, D.C.: Board of Governors of the Federal Reserve System (March)

Kelly PM, Adger WN (2000) Theory and Practice in Assessing Vulnerability to Climate Change and Facilitating Adaptation. Climate Change 47

Leichenko RM, O'Brien KL (2001) The Dynamics of Rural Vulnerability to Global Change: the Case of Southern Africa. Mittigation and Adaptation Strategies for Global Change 7. Kluwer Academic Publishers. Netherlands

Lewis WA (1954) Development with Unlimited Supplies of Labour. Studies, Manchester School of Economics and Social

Liverman D (1990) Vulnerability to Global Environmental Change. In: Understanding Global Environmental Change, Ed. Kasperson. Annals of the Association of American Geographers

Marx K (1867) Expropriation des Landsvolks von Grund und Boden. Die so genannte Ursprüngliche Akkumulation. 24. Kapiteln. In: Kritik der Politischen Ökonomie. Erster Band. Nach der ersten von Friedrich Engels herausgegeben Auflage. Hamburg, 1864

Montserrat H, Chavez M (2003) Tres Modelos de Política Económica en México durante los últimos sesenta años. In: Análisis Económico No. 37., Vol. XVIII. Universidad Autónoma Metropolitana. Mexico City

Norris FH, Stevens SP, Pfefferbaum B et al (2008) Community resilience as a Metaphor, Theory, Set of Capacities and Strategy for Disasters Readyness. American Journal of Community Psychology 41:127–150

Oliver ML, Shapiro TM (1990) Wealth of a nation: a reassessment of asset inequality in America shows at least one third of households are asset-poor. American Journal of Economics and Sociology 49(2)

Prebisch R (1973) La cooperación internacional en la política de desarrollo latinoamericano. Santiago

Prebisch (1951) Problemas teóricos y prácticos del crecimiento económico. Santiago

Prebisch (1950) The economic development of Latin America and its principal problems. United Nations, New York

Ravillon M, Datt G (1996) How important to India's poor is the Sectoral Composition of Economic Growth? World Bank. Economic Review. Vol. 10 (1)

Ricardo D (1817) Chapter XXI Effects of Accumulation on Profits and Interest. Principles of Political Economy. Third Edition. Everyman Library, London

Rodríguez O (1980) La teoría del subdesarrollo de la CEPAL (The ECLAC's Underdevelopment Theory). Ed. Siglo XXI, Mexico, City

Sarris A (2001) The Role of Agriculture in Economic Development and Poverty Reduction: an Empirical and Conceptual Foundation. Rural Development Department. The World Bank Group. Rural Strategy Background Paper No. 2. Washington, DC

Schjolden A (2003) Are Vulnerability and Adaptability two sides of the same coin? A critical view of climate impacts research. In: Update IHDP, 04/2003. Bonn, Germany

SEDESOL (2002) Discurso de la Secretaria Josefina Vazquez Mota. Serie documentos de Investigación. Secretaría de Desarrollo Social, Mexico City

Sen A (1981) Poverty and famines: An essay on entitlement and deprivation. Oxford University Press, Oxford

Sherraden M (1991) Assets and the poor. M. E. Sharpe Inc., Armonk, New York

Slesnick DT (1993) Gaining ground: Poverty in the postwar United States. Journal of Political Economy 101:1–38.

Singer H (1950) The distribution of gains between borrowing and investing countries. American Economic Review 40:473–485

Tejo P (2000) La pobreza rural. Una preocupación permanente en el pensamiento de la CEPAL. Red de Desarrollo Productivo. Unidad de Desarrollo Agrícola. División de Desarrollo Productivo y Empresarial. ECLAC. Santiago. http://www.eclac.cl/publicaciones/DesarrolloProductivo/4/lcl1454/LCL1454P.pdf

Timmer CP (1997) How well do the Poor connect to the Growth Process? CAER Disscussion Paper No. 178. Harvard Institute for International Development. Cambridge, Massachusetts. The MIT Press

Todaro MP (2000) Urbanization and Rural-Urban Migration: Theory and Practice. Economic Development. Seventh Edition. Addison Wesley Longman. Essex, In

Todaro MP (1982) Economics for a Developing World, 2nd edn. Ed. Longman Group Limited, Essex, UK

Turner BL et al (2003) A Framework for Vulnerability Analysis in Sustainability Science. 8074–8079 PNAS. July 8, 2003. Vol. 100. No. 14. www.pnas.org/cgi/doi/10.1073/pnas.1231335100

UNCTAD (2003) Economic Vulnerability: the UN approach to the economic vulnerability of Least Developing Countries: implications for Asian and Pacific LDCs. United Nations Conference on Trade and Development. Pp. 2-7. Bangkok

Vatsa KS, Krimgold F (2000) In: Managing Disaster Risk in Emerging Economies; "Financing Disaster Mitigation for the Poor". The World Bank

World Bank (2004) Poverty in Mexico: an assessment of conditions, trends, and government strategy. Report No: 28612. Mexico City

World Bank (2002) Country assistance strategy of the World Bank Group for the Mexican United States, 2002. Washington, DC

World Bank (2001a) World Bank Development Indicators. World Bank, Washington, DC

World Bank (2001b) Poverty Reduction in Indonesia: constructing a new strategy. Report No. 23028-IND. Washington, DC

Wolff E (2001) Recent trends in wealth ownership, 1983-1998. In Shapiro and Wolff (eds.), Assets of the Poor, Russell Sage Foundation. New York

Yap JT (2002) Monitoring Economic Vulnerability and Performance: Applications to the Philippines. Philippine Institute for Development Studies, Thailand Development Research Institute, Manila

Zapata-Martí R, Saldaña-Zorrilla SO (2009) Evolución de la Epidemia de Dengue en el Estado Plurinacional de Bolivia en 2009. Comisión Económica para América Latina y el Caribe (CEPAL), Naciones Unidas y Organización Panamericana de la Salud (OPS). 65 pp. March 2010. Santiago, Chile

Chapter 3
Natural hazards and economic stressors

No cabe duda que es preciso superar el liberalismo del siglo XIX.

ORTEGA Y GASSET, La rebelión de las masas

Abstract This chapter explains the mechanism by means of which natural disasters and some past economic policy decisions have turned into hazards in Mexico. Natural disasters occurrence is increasingly producing severe damages to the so-called traditional agriculture, highly exposed to climatic events due to its predominating rainfed cropping practices as well as its high marginalization conditions, which together tend to amplify the negative effects from hazards. In the frame of the economic reforms implemented from the middle of the eighties, trade liberalization has led some economic sectors to increase more remarkably their exposure to international markets. The negative impact of trade liberalization on rural livelihoods has been evidenced over the past two decades through price drops of agricultural grains, the main crop of subsistence farmers. It has undermined their incomes given their limitations to increase neither productivity nor cropping land, as well as their inability to re-orientate production.

Keywords Hazards • Exposure • Weather-related disasters • Geologic disasters • Trade liberalization • Rural livelihoods • Foreign trade • Agricultural policy • Rural poverty • Rural development • Agricultural productivity

To explain the mechanism by means of which natural disasters and some past economic policy decisions have turned into hazards in Mexico, Section 3.1 describes the historical and spatial distribution of economic losses from natural hazards in Mexico[1]; Section 3.2 analyzes the specific weight of natural hazards in the agricultural sector, further discussed in Section 3.3, which assesses its influence on the poor and on rural-urban emigration in the country[2]; Section 3.4 discusses the contribution of trade liberalization to increasing economic vulnerability in Mexico, and Section 3.5 concludes presenting a balance of these hazards in shaping economic vulnerability.

---

[1] It is analyzed in more detail in Chapter 4, which models it with coping- and adaptability-related variables.

[2] More deeply analyzed in Chapter 5.

© The Author 2015   25
S.O. Saldaña-Zorrilla, *Natural Disasters, Foreign Trade and Agriculture in Mexico*,
SpringerBriefs in Environmental Science, DOI 10.1007/978-3-319-17359-7_3

3.1   Historical losses from natural hazards

Like Japan and Central America, the Pacific Cost of Mexico is located at the so-called *Circum-Pacific-ring of fire* of tectonic activity. Hurricanes are a regular hazard on both the Atlantic and Pacific Coasts. Roughly speaking, the South of the country suffers high floods frequency, whereas the North of droughts and of an ongoing desertification process. Over the past four decades, natural disasters in Mexico have increased economic cost (see Fig. 3.1 below).

However, most of that affectation has been due to weather-related disasters, responsible for ca. 80% of economic losses over the period 1980-2005 (see Table 3.1). As observed in the next section, the weight of hurricanes in total disasters losses is particularly remarkable as it has implied 68% of total losses over the same period.

3.1.1   Hurricane

For its geographic location, Mexico can be hit at the same time by two independent cyclones, namely from the North Atlantic and the North Pacific. Alone between the period 1967-1977 Mexico suffered 57 hurricanes, which caused severe damages particularly in six states in the South of the country (CENAPRED 2001).

These hurricanes have mostly originated in the Caribbean Sea and the Central American Pacific Cost, moving into Mexican territory. Hurricane Paul in 1982 moved over the Pacific from El Salvador and Guatemala, upgraded from a tropical

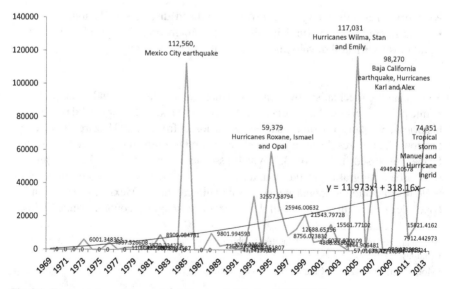

Fig. 3.1   Economic losses from natural disasters in Mexico (1969-2013)

Table 3.1 Losses from disasters in Mexico 1980-2005 (Losses in current USD mill)

Disaster type	Direct losses	Indirect losses	Total losses	As % of total
Weather	21,887	145	22,032	79
Geologic	4,044	517	4,561	16
Human	1,150	134	1,284	5
Total	27,081	796	27,877	100

With data from Guy Carpenter (2006) and CENAPRED (2001)

Table 3.2 Major hurricanes in Mexico, 1980-2005 (losses in current USD mill)

	Hurricane	Economic losses*
1982	Paul	82
1988	Gilbert	597
1990	Diana	91
1993	Gert	114
1995	Opal	151
1997	Paulina	448
2002	Isidore**	235
2005	Emily	250
2005	Wilma	15,000
2005	Stan	1,500
Losses 1980-2000		18,968

Period 2002-2005: data from Guy Carpenter years 2003 and 2006 (2006); years 1980-2001 (CENAPRED 2001)
*Period 1980-2000 are estimates of losses to exclusively public assets.
**Losses estimated at 50% total economic losses reported from Guy Carpenter in Mexico and the USA from Isidore, and in Central America and Mexico from Stan.

storm in these countries[3] to a 100 mph hurricane once in Mexican territory.[4] Hurricane Pauline in 1997 caused floods and mudslides, whipping up strong waves and currents with devastating effects across the coastline of Oaxaca and Guerrero. Alone in Acapulco it killed over 200 people and caused over US$400 mill to public assets (see Table 3.2). This region's uneven topography contributed to flooding and landslides following the hurricane. This calamity evidenced the rapid and wrong

---

[3] Over 1,000 fatal victims in Central America.

[4] The Southeast of Mexico shares with Central America and Caribbean countries not only the above presented high hurricanes and earthquakes affectation, but also the fact that disaster-vulnerable population is highly poor. Affected population in Mexico are usually located in southeast states, which, like in Central American and Caribbean countries, are subsistence farmers without access to credit, low crop insurance coverage, and work small farms. However, most the times these countries are hit by the same hurricane, relative higher economic losses and calamities occur in Central America and the Caribbean compared to Mexico. It is so particularly due to differences in coping and adaptive capacity among them. It suggests the potential applicability of this thesis approach of vulnerability reduction to these other countries.

projected urbanization process in large costal cities of the recent years, which has led to increase potential economic and human losses from extreme climatic events.

Munich Re said the 2005 hurricane season to be the most costly of all time (Munich Re 2005). In turn, Mexico was severely damaged by hurricanes in that season, particularly on the Peninsula of Yucatan, where hurricanes Emily, Wilma and Stan hit within 3 months. Damages from Emily and Wilma accrued to mainly the tourist sector, whereas Stan hit mainly the poor in both urban and rural areas.

Emily made landfall on the Yucatan Peninsula on 18 July, provoking the evacuation of around 130,000 tourists from luxury beachfront hotels in the resort city of Cancun. The storm produced heavy rains and winds, demolishing buildings, triggering floods and affecting especially tourism. Almost 3 months after, hurricane Wilma destroyed thousands of homes and hotels and flooded luxury hotels and resorts (Guy Carpenter 2006), whose economic losses were fivefold higher than those from Emily, becoming Mexico's most costly hurricane ever (see Table 3.2).

In October 2005, hurricane Stan passed over the Peninsula of Yucatan, but derived climatic events caused dramatic human and economic losses hundreds of kilometers to the South in especially marginalized areas. Hurricane Stan itself was not as damaging as its derived strong winds, floodings and mudslides on Chiapas, Oaxaca and Veracruz, killing 42 people and destroying assets of over one million people, exceeding USD 1.5 billion in direct losses. In Guatemala the magnitude of Stan was higher in terms of human fatalities (1,400 dead), but lesser in economic losses, ca. USD 1 billion.[5]

3.1.2 Floods

Whereas flooding caused by rivers overflowing their banks occurs almost every year in Mexico (especially in summer time), flooding from heavy rainfall occur permanently, causing erosion, landslides, and severe damage to housing, agriculture, livestock and public infrastructure. This section comprises river overflowing and heavy rainfalls only as they cause flooding. Flooding is naturally less frequent in semi-arid regions, but however it can be particularly devastating when it does occur there.[6]

Heavy rains alone rarely cause floods. Soil erosion caused by deforestation, inadequate agricultural practices, and increasing urbanization contributes to increasing flood risk (see Conde and Saldaña-Zorrilla 2008). Despite considerable invest in drainage infrastructure, Mexico City experiences land losses every year from flash flooding.

The occurrence of flooding is increasing in recently urbanized plain areas as change in land use is broadening the capture of rainfall, producing flows that the natural basin cannot cope with. The loss from flooding in Mexico has been high.

---

[5] However, losses in Mexico compared to national GDP are equivalent to 0.002%, while in Guatemala it reached 3.5% of GDP.

[6] As in the arid region of Monterrey in 1998 (CENAPRED 2001).

Alone between 1970 and 1990 more than 1,800 people died and about USD $7.7 billion in economic losses are estimated to be incurred from over 1,000 floods events in Mexico.[7]

In September 1998, floods and landslides struck Chiapas, turning into the worst disaster in Mexico after the 1985 earthquake. It devastated huge extensions in coastal areas, and left some 800,000 people homeless (25% of Chiapas population), and caused more than 200 deaths. The most immediate problem was to reach the (isolated) areas affected for the flooding to provide aid relief and reconstruction to rehabilitate destroyed transport and telecommunication infrastructure. Other Mexican states affected by the torrential rain were Guerrero, Morelos, Oaxaca, Guanajuato and Nayarit, and they are also considered highly prone zones to landslides.

3.1.3   Earthquake

Historically, a number of large earthquakes have occurred within the Trans-Mexico Volcanic Belt, located approximately at 20° N latitude. This area runs from west to east, and its high tectonic activity is accompanied of active volcanic and faulting. Some of the largest earthquakes in recent history –like the 8.1 and 8.0 Richter degrees Michoacan and Colima earthquakes in 1985 and 1995, respectively- originated in this region and spread across the pacific cost (Currie et al 2002).

Most seismic activity in Mexico is related to the active tectonic boundaries between the North America and the Pacific and Cocos plates (Currie et al 2002; Bitrán et al. 2001). A northern boundary that accommodates friction between the North America and Pacific plates runs beneath the Gulf of California and runs southward parallel to the Pacific cost of Mexico (World Bank et al 2000). This area is part of the called *circum-Pacific Ring of Fire*, which draws a circle along Eastern Oceania's Islands, Asian Pacific, and North and Central American Pacific costs.

Other significant earthquakes in the recent times have been originated in this area as well, like in January 30, 1973 (magnitude 7.5), November 29, 1978 (magnitude 7.8), March 14, 1979 (magnitude 7.6) and September 19, 1985 (magnitude 8.1) (CENAPRED 2001). In general, these earthquakes' epicenters are located on seismic gaps, that is, on places where at least one strong earthquake has stunned in the past, but where no earthquake has taken place for a long time (World Bank et al 2000).

As observed in Fig. 3.2, CENAPRED has elaborated a Seismic Map of Mexico based on historic registers of major earthquakes in Mexico and on data of terrain acceleration from major earthquakes. The map shows four zones: Zone A represents areas without seismic activity over the last 80 years, which terrain acceleration is expected to keep fewer than 10%. Zone D is the area with high earthquake frequency, and its acceleration exceeds 70%. Zones B and C are mid-intensity regions.

---

[7] With data from La Red.

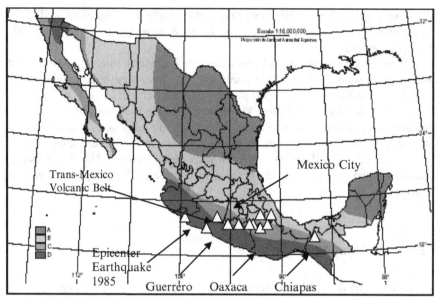

SOURCE: Saldaña-Zorrilla 2007.

Fig. 3.2 Seismic map of Mexico. Primary source: CENAPRED, Ministry of the Interior of Mexico

Table 3.3 Earthquakes in Mexico 1980-2014 (losses in current USD mill)

Earthquake	Dead	Direct losses	Indirect losses	Total losses
Mexico City 1985	6,000	3,589	515	4,104
Colima 1995	0	21.1	-	21
Oaxaca, Puebla and Morelos, June 1999	15	151	0	151
Oaxaca, September 1999	35	153.6	1.4	155
Guerrero 2001	0	2.9	-	3

Data from CENAPRED (2001), Ministry of the Interior, Mexico.

In 1985, the *Michoacan Earthquake* caused severe damage and collapse of high-rise construction in Mexico City (located on Zone B) even despite its long distance from the epicenter. It is said to be so due to the fact that the period of vibration of deep lacustrine soils beneath Mexico City coincided with the fundamental period of many high-rise buildings (Singh and Suárez 1987). This earthquake produced US$ 4, 104 in economic losses, the most costly earthquake during the analyzed period (see Table 3.3). This event generated relevant social and politic changes in some cases, and accelerated other existing ones (see box 1). Derived from a natural hazards survey from the World Bank, IIASA and Guy Carpenter of 2000, Mexico can expect a 6.5 Richter degree or greater earthquakes every two years, a magnitude 7.0 or greater about every two years, and a magnitude 8.0 or greater about every 33 years (World Bank et al 2000).

Box I: Political changes after the 1985 Earthquake

The 1985 Michoacan earthquake in Mexico City caused 89% of total geologic losses over the period 1980-1999 (over 4 US billion, ECLAC, CENAPRED 2001), and is considered a break-point in Mexico's recent history. Some authors maintain that this amount could be imprecise, since original estimations of the Mexican government overinflated damage figures up to $US6bn as a lever to relax IMF conditions for reconstruction loans (Albalá-Bertrand 1993:140, Castillo 1985). ECLAC estimates were more conservative (US$ 4.1bn.), but it was estimated based on official loss assessments. By contrary, these authors presume the number of deaths to have been underestimated in order to avoid the army undertaking emergency control, as required by Mexican law* (Castillo 1985). After the students' genocide in 1968 in the Three Cultures Square in Mexico City, military intervention was believed to irritate civil society in Mexico, which had added tension to the disaster.

This earthquake undermined the image of the Federal government as warrant of national security in the country due to the delay and badly coordinated response (Monsiváis 1987), as well as due to the governmental corruption evidenced through constructions approved by the authority, despite not fulfilling building codes requirement (Castillo 1985; Ramírez 2005). Since this disaster, corruption started to be perceived country wide as beyond the economic burden and the public moral issue it is, but also as a factor of risk to citizens' life. The insufficient governmental preparedness and response to this disaster led the civil society to undertake rescue, cleaning up, and relief operations (Ramírez 2005; Monsiváis 1987). It generated a collective feeling of an absent state and led to crucial political transformations.

The political cost of such insufficient response was paid in the upcoming national elections in 1988. Most likely influenced by these events, the hegemonic PRI - the ruling party since 1929-, split. Corriente Democratica, an intern organization of PRI, joined the leftist PSUM party, which obtained a gigantic electoral triumph in 1988, including dozens of seats in the federal congress and municipal governments (Calderón and Cazés 1991), and perhaps the presidency of the republic as well. PRI is widespread suspected of committed electoral fraud through the electronic votes counting system (Acosta 2006). In any case, that political change proved to have contributed to the fast electoral growth of the opposition in the oncoming years as the congress reached a real multiparty representation six years later, leaving behind the one-party political system of the previous six decades.

*Plan DN-III-E, Art. 42, Ley Orgánica del Ejército y Fuerza Aérea Mexicanos.

3.1.3.1 Natural-social stunning events

Over 50% of Zone A (most prone to earthquakes) is located along Mexico's poorest states: Oaxaca, Chiapas and Guerrero (see Fig. 3.2), recurrently hit by hurricanes and floods as well. These states are historically characterized by having Mexico's lowest *per capita* income, highest illiteracy rates, and worst health facilities in Mexico. Coincidently, all Guerrilla movements after the Mexican Revolution of 1910 have popped up in precisely these three states, i.e. *Lucio Cabañas* Guerrilla in Guerrero during the 1970s, EZLN *Ejercito Zapatista de Liberacion Nacional* in Chiapas since 1994, and EPR *Ejercito Popular Revolucionario* and ERPI *Ejercito Revolucionario del Pueblo Insurgente* in Guerrero and Oaxaca from 1995.

3.1.4   Volcanoes

As a result of the collision between the North America and the Pacific and Cocos plates, dozens of volcanoes are distributed along Mexican territory, but especially on the Trans-Mexico Volcanic Belt (*Cordillera Neovolcánica,* see 7), which defines the geomorphologic boundary between North and Central America.[8] In fact, 14 of these volcanoes have erupted in recent history.

3.1.4.1   Chichonal

The most recent volcano eruption that resulted in catastrophic losses in Mexico was *El Chichonal* (State of Chiapas) in 1982, which completely destroyed eight communities and claimed 2,000 lives. The eruption lasted nearly six hours and caused severe damages to crops, cattle, and cultivable land within a 50-kilometer radius (CENAPRED 2001).

3.1.4.2   Popocatepetl

Between 1993 and 1998, volcanoes *Popocatepetl, Colima, Tacaná,* and *Everman* heve registered activity. According to seismographers from the National Autonomous University of Mexico (UNAM), a major eruption of *Popocatepetl* would affect thousands of people in settlements within a 20-30 kilometer radius (CENAPRED 2001). This volcano is being carefully monitored for preparedness sake, including evacuation measures for communities in the vulnerable area.[9]

---

[8] In geomorphologic terms, the Trans-Mexico Volcanic Belt defines the boundary between the North American *Rocky Mountains-Sierra Madre Occidental* system, and the Mesoamerican *Sierra Madre del Sur- Sierra Madre de Chiapas.*

[9] In large cities –such as Puebla, Cuernavaca, and Mexico City, over 40 kilometers away from *Popocatepetl-*, the effects of an eruption are likely to be limited to falling volcanic ash.

3.1.5   Droughts

In Mexico only around 20% of harvested hectares is irrigated (ECLAC 2006; INEGI 2005). This fact reveals a high dependence of agricultural production from meteorological phenomena. Lack of water in form of droughts, forest fires, and high temperatures damaged over 36 million hectares of crops over the period 1970-2002 (García and De la Parra 2002), which means 18 times the total cultivated surface in Mexico in 2002. Alone in the 1980s, economic losses from droughts reached US$1.2 billion (CENAPRED 2001).

The graphic below shows the distribution of 6, 296 disasters (natural and anthropogenic) in Mexico that occurred over the period 1970-2001. Over one third of them are events caused by excess of water (floods, heavy rains, storms, hurricane, etc., 2,208 reports), whereas lack of water represented 18% of total events. Floods are the most frequently reported event (blue bar, over 1,112 reports), followed by fires (828 reports) and frosts (407).[10]

Although other natural disasters have damaged less spectacularly, they have been more frequent.[11] Frost is the kind of phenomena which have damaged agriculture more permanently, accumulating MX$95, 910 mill over the period 1979-1988, followed by hailstorms (MX$30,153 mill) (CENAPRED 2001).

3.2   Hazard exposure of agriculture

Macroeconomic analyses tend to ignore crucial impacts of natural disasters on the economy. Though the agricultural sector in Mexico is small in GDP terms (4%, compared to 68% from services, 28% from industry[12]), this sector employs 68% of the population in extreme and moderate poverty in rural areas.[13] Moreover, the Mexican agriculture is highly vulnerable to weather-related disasters given current levels of rainfed agriculture. The 82 major weather-related events that have struck between 1980 and 2002 have damaged mostly agriculture disrupting rural incomes, as discussed throughout this section (Fig. 3.3).

---

[10] This data basis has been created by Desinventar (DesenRedando, La Red 2004) by collecting reports from media, especially from three Newspapers for the Mexican case: El Universal, La Jornada, and Excelsior. Given that fact, data arising from Desinventar must to be carefully employed, since in detailed analysis some data may be incomplete, leading to either over- or underestimated losses due to its journalist nature. However, it is helpful to get general pictures of disasters. It also provides good approaches on magnitudes and disaster frequency. Amount of losses are still controversial, since they vary widely from observation to observation, and rarely coincide with other sources.

[11] In fact, most of the increase in weather-related disasters may be linked to El Niño phenomena (Vatsa and Krimgold 2000: 131; CENAPRED 2001: 145)

[12] Instituto Nacional de Estadistica, Geografia e Informatica (INEGI), Aguascalientes, Mexico.

[13] Country assistance strategy of the World Bank Group for the Mexican United States (World Bank 2002).

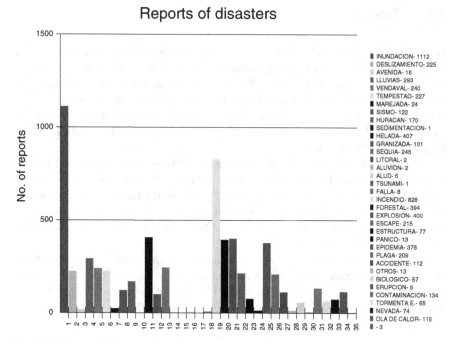

Fig. 3.3 Disasters in Mexico 1970-2000 by kind of event

Every time a hurricane strikes, over 70% of total damages are located in agriculture. Thanks to reports from La Red (2004), we know that natural disasters affectation in agriculture over the past 35 years has affected mostly assets of poor and extreme poor farmers.

- In 1982, Hurricane Paul and floods caused MX$7.4 millions in direct losses to agriculture, equal to 70% of total losses from this event.
- Floods in 1985 hit Mexico, and 85% of total losses were located in agriculture, damaging especially crops. In the same year, heavy rains caused losses at MX$4,177 mill, 97% in agriculture (infrastructure and crops).
- In 1988, Hurricane Gilbert hit severely the Peninsula of Yucatan and Northeastern Mexico, damaging the whole local economy. Losses in the agriculture sector were especially acute, since it meant 86% of total losses (US $ 65 mill.). Along the same year, three more hurricanes hit Mexico causing economic losses to crops three fold Gilbert's losses (CNA 1998; CENAPRED 2001).
- Hurricane Paulina hit Oaxaca and Chiapas (the poorest Mexican states) in 1997, sparking off severe damages on the whole economic activity and housing, but 88% of total losses are in agriculture and livestock.
- Agricultural losses from climatic events repeated in 1998 in these two states again, though more dramatically in Chiapas. Heavy rains caused landslides and avalanches from the southern mountains of Chiapas to the cost. This disaster

reflects differences in living conditions within population in these countries as in both countries assets of lower income people were more affected. Additional evidence in the same way is found in Honduras with hurricane Mitch (Vatsa and Krimgold 2000), El Salvador Earthquake in 1986 (ECLAC 1986), Dominican Republic with hurricane Georges (Butterfield 1998), the United States of America when hurricane Kathrina hit in 2005 (O'Brien et al. 2005), among others.

As discussed above, hurricanes Wilma and Stan affected the Yucatan Peninsula and Chiapas in 2005, with relatively higher damage to assets of the poor. Economic losses from Wilma exceeded USD 15,000 mill, while those from Stan in Chiapas were at USD 3 mill. However, the adaptive capacity of affected population from Wilma in Yucatan proved to be much higher, where over 50% were insured losses (Guy Carpenter 2006). It was so especially due to the fact that most losses from Wilma took place in Cancun, damaging luxury hotels and resorts. By contrast, losses derived from Stan were totally uninsured, affecting basically assets of the poor both in marginalized slums in urban areas (i.e. Las Americas in Tapachula) and in subsistence farmers regions (i.e. Escuintla, Mapastepec, Cacahoatan). Facts like that provide the motivation to assess coping and adaptive capacity in that region, presented in Chapter 6 derived from a fieldwork of the author to these municipalities in 2004 and 2014.

The map below plots 1,372 reported events on agriculture affectation in Mexico over the period 1970-2001. The map is divided in states, and colors represent ranges of damages. As observed, the most damaged state is Chiapas, followed by Oaxaca and Yucatan (in the South), Mexico and Guanajuato (Center), and Jalisco, Tamaulipas, Sonora and Sinaloa (North). Chapter 5 overlaps disaster affectation maps with distribution maps of income, marginalization and access to risk management instruments, showing that the South is relatively the most vulnerable region in the country.

At a first sight, economic losses from natural disasters seem to be similar across the national territory, but it produced dramatic effects on marginalized regions without access to risk management instruments in mostly rural areas. Though the 1985 Michoacan earthquake is considered Mexico's all time most dramatic earthquake -losses reached 2.17% of GDP in year 1985-the huge differences in coping and adaptive capacity between Mexico City and Chiapas has made the latter to absorb much more negatively past hazards. The 2003 poverty report issued by the Mexican government points out that Chiapas is now the poorest state of Mexico, both in terms of GDP per capita as well as in social indicators. In addition, the development gap between Chiapas and the other states of Mexico is continuously enlarging. Thus, the relevance of natural disasters lies on the fact that they do not hit significantly upon the whole economy, but to the agricultural sector in highly marginalized regions with low coping and adaptive capacity. These hypotheses are quantitatively tested through the spatial econometric model presented in this book (Fig. 3.4).

Given an adverse entitlements system and assets distribution, natural disasters affectation in the Mexican countryside tends to exceed rural coping and adaptive capacity and so to terminate rural livelihoods, triggering domestic rural-urban and broad emigration. As an efficient strategy for agriculture and rural areas was not

buried dozens of communities, destroyed 712 Km of asphalted roads an
50% of rural roads of the state (3, 600 Km.), 22 bridges, etc. Over 25% c
infrastructure of Chiapas was totally destroyed. Losses reached US $ 60
too high for a poor state.[14]

- By 1999, heavy rains continued damaging agriculture in Southern Mexico. S
  poor communities in the states of Puebla, Veracruz and Tabasco were so
  affected by heavy rains. Inhabitants of these communities survive historic
  agricultural activities, and 66% of total losses from these heavy rains acc
  rural infrastructure and crops of small-scale producers (MX$1,767 mill.).

3.3 Relative vulnerability of the poor

Though being poor does not necessarily imply being vulnerable, poverty
individuals relatively more vulnerable to a given hazard. People worldwide
adverse economic conditions are less able to invest in all items, including tl
manage risk and increase disaster protection. Developing countries have his
been more severely damaged compared to developed countries (Norris et
Benson and Clay 2002). On the one hand, total economic losses tend to be
rich countries in absolute terms, but compared to economy value, losses a
higher in developing countries (i.e. Freeman and Mechler 2001). A give
hazard with identical intensity can hit in different degree two distinct c
Differences in civil protection system, health facilities and public financi
(i.e. for reconstruction) cause countries to absorb hazards differently. For
the same hurricane hit the Dominican Republic and Haiti in 2004, but
economic losses fivefold higher in Haiti. Though both are developing (
they are at different development stages. As Cannon (1994: 24-26) points
turns a natural hazard into a disaster is not simply a question of money, b
economic and political system. The way countries structure societies d
that similar hazards lead to very different impacts among societies.

Given current entitlements, the poor is the stratus most prone to suffer f
ral disasters, especially in developing countries. The distribution of humai
many developing countries reveals high inequity. The most productive
terrains belong to middle- and upper classes, whereas less productive and
areas were left to the poor. Most of the victims of Guatemala's Earthqual
were poor (23,000 deaths), who lived in ravines and gorges, areas very
disaster in case of earthquake or landslides. The river Oder, which divides
from Poland, overflowed in 1997 producing severe floods. Lack of maint
dykes and flood defenses, together with poor people living along the ri
polish side, produced disgracefully notoriously higher damages there t
German side (Vatsa and Krimgold 2000). That reveals, on the one hand,
differences to mitigate disasters between these countries. On the othe

---

[14] Losses reached 9.3% of Chiapas GDP in 1998.

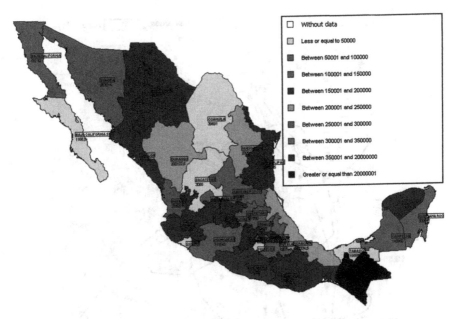

Fig. 3.4 Damaged hectares due to natural disasters in Mexico by state (1970-2000)

successfully undertaken during the 80's –discussed in Chapter 2 and further in Chapters 3 and 4-, these rural workers have been migrating massively to especially large domestic cities, i.e. Mexico City, Monterrey, Guadalajara, and Tijuana, building irregular settlements (mostly slums) on areas very prone to natural disasters as well. Out migration from Mexico to other countries has increased over the past two decades. The Population Census of 1990 reports that 0.24% of Mexican population was residing abroad, whereas in the 2000 Census this figure rose to 0.41% (INEGI 2005).

In Mexico City, the most disaster-prone areas are populated by the poor. Socioeconomic maps of Mexico City show the poor located along an area going from Southeast to Northwest, wide in the South and narrower moving northwards (INEGI 2006), exactly on very prone areas to natural disasters, whereas richest areas run from Southeast to Northeast, that is, on the less prone areas to disasters.[15] Fig. 3.5 below relates zones at high risk to earthquake in Mexico City and slums location. Violet color represents areas whose underneath is the lake Tenochtitlán.

The severe damage and collapse historically experimented in this area has occurred due to the accelerating vibration of deep soils beneath (resonance effect), as in 1985 when the Michoacan earthquake hit Mexico City. In red circles

---

[15] Experts on demography used to name this socio-economic distribution as The Bell of Income, due to its geographic shape: poor from south-west to the north, each time narrower; rich from north to south-east, each time wider; and a mixture of middle, high and low income in the middle.

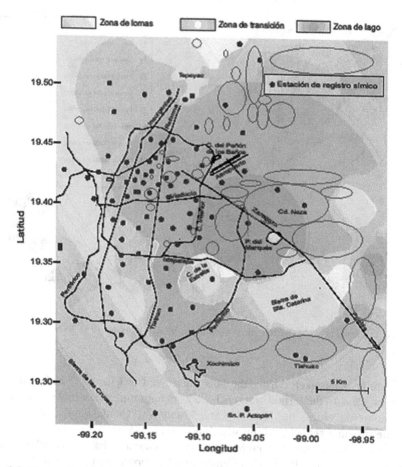

Fig. 3.5 Earthquake risk in Mexico City and slums distribution

are represented slums and very poor areas, which are mostly located upon this high-risk area.

The nature of the Mexican economy is dualistic, since in general terms there exists a modern urban capitalist sector geared toward capital-intensive large-scale production, coexisting with a traditional rural subsistence sector geared toward labor-intensive small-scale production. In the same way, this dualism is observed in the urban economy, which is divided into formal and informal sectors. In Mexico, estimated share of urban labor force in the informal sector is 57%, and slums as percentage of population in Mexico City is reckoned to be 46 (ECLAC 2003). This city concentrates over 20 million people,[16] the world's second largest city in terms

---

[16] 15.6 million people live alone in Distrito Federal, plus 5.1 in neighbour municipalities, which have been integrated into the metropolis. Sources: World Resources Institute 1996-1997: The urban environmental (New York: Oxford University Press) tab. 1.1; United Nations, World Urbanization Prospects. See Appendix B for income distribution.

of population (after Tokyo, 26 mill.), and the world's largest in terms of surface ($4,986km^2$) –INEGI 2005.

About one-quarter of Mexico's labor force is still employed in agriculture, and as conditions in the countryside continue to worsen, rural workers do not have incentives to stop migration not only to urban Mexico, but also to urban and rural USA (Robinson et al 1995). Mexico is the first country of origin of migrants to the USA, where nowadays 1 out of 3 migrants was born in Mexico, integrating the first majority migrating community in the USA as well as in 31 of the 50 states of the American Union (Center of Migration Studies of Washington 2003, US Census Bureau 2002).

3.4   Trade liberalization

In previous sections we discussed the relevance of natural hazards to the agricultural sector and in turn to the rural poor in Mexico. This section discusses basic concepts leading to understand why trade liberalization turns into a hazard despite its high expectations of becoming a source of benefits for even subsistence farmers. It presents, as well, empirical evidence of the increasing imports trend and decreasing agricultural prices undermining rural incomes.

3.4.1   Globalization and regionalization

Before starting to explain trade liberalization, let us clarify some key under-lying concepts, like globalization and regional markets. The idea of globalization can be first understood by simply language interpretation. Global is an adjective implying the earth globe. The dimensions of globalization are still polemic. Some authors opt to include only visible measurable aspects into the globalization concept, like flows of trade, capital and people across the globe.[17] Others denote in addition a stretching of social relations and activities across regions and frontiers, implying cultural and symbolic dimensions (Held and McGrew 2000). In turn, these broader concepts used to include achievements of institutions and organizations of the international community dealing with global shared knowledge (i.e. UNESCO), international aid (i.e. Red Cross, Medecins Sans Frontiers, etc.), and even going beyond by including scopes from rather modern multilateralism (i.e. UNO, NATO, APEC, etc.). However, we prefer to limit this concept to just include trade, capital and people for

---

[17] A frequent questioning of globalization is about its innovation. Helleiner (1997) considers globalization to be not only a phenomenon of the modern age, but placed in the context of a secular trend of world historical development. So, one can find similar globalization patterns connecting the second half of the 19th century (*Victorian era* of the English empire, *la belle epoche* in France), with the Spanish colonialism in the centuries 17th and 18th (i.e. *Galeon Manila-Acapulco-Cadiz*) and even with the ancient Silk Route.

this work's sake.[18] Some arguments criticize the spatial validity of denominating *globalization* the current stage of world economic integration, opting instead to call it *regionalization*. Castells (1996) points out that the world is rather passing by an increase in interregional relations. In terms of merely trade, the case of Mexico is more in line with that view.

Despite the current trade agreements between Mexico and over 40 countries, Mexican foreign trade continues very concentrated with USA. These trade agreements include the North America Free Trade Agreement (NAFTA), the Mexico-European Union Free Trade Agreement (MEUFTA), and the Mexico-Central America Free Trade Agreement (MCAFTA), among others. In addition, Mexico is a member of the World Trade Organization (WTO) since 1986 (GATT that time) and is preparing negotiations for expanding trade agreements to the rest of the American continent by means of the FTAA (Free Trade Agreement of the Americas). For that reason, Mexico is usually considered one of the most opened world economies. However, ca. 90% of Mexican foreign trade value is concentrated with exclusively the USA (Saldaña-Zorrilla 2003). Beyond trade, other productive factors are highly concentrated with the USA as well, i.e. labor move[19] and foreign direct investment.[20] That fact is understandable if one looks at the relevance of geographic aspects in factors mobility, which in turn seems to be the main limit of Mexico for going global.

3.4.2   Growing share of foreign trade in the economy

As the share value of foreign trade compared to the whole value of the economy (GDP) increases, economies like the Mexican become more exposed to external factors. Foreign trade expansion is conventionally said to have a positive effect on the economy as a whole (Frankel and Romer 2000), especially when boosting exports (Micco and Perez 2001). Also, when foreign trade grows faster than the sum of the other components of the economy (consumption, investment and public sector expenditure), foreign trade becomes relatively more important in a weighted

---

[18] A conventionally admitted aspect of globalization is its growing magnitude or intensity of global flows such that states and societies become increasingly inserted in world systems and in networks of interaction based on continuously improving communications facilities.

[19] In terms of flows of people (legal and illegal), the trend is very similar too. Unlike some negligible programs on temporary work, Mexico has not signed trans-boundary labor agreements with the USA, and despite that fact, over 90% of international emigration goes to that country.

[20] Quiroz (2002) reckons that 60% of foreign direct investment in Mexico arose from the USA in 1995 –the rest from Germany (5%), Japan (4.5%), etc. There is widespread criticism of the fact that most FDI in Mexico is comprised of investments of multinational companies in their existing branches in Mexico, rather than of new business implementation and incorporating new economic agents to the benefits of foreign trade (Dussel 2000). In addition, capital and exports concentration in very few hands is taking place following trade liberalization in this country (Saldaña-Zorrilla 2003).

appreciation of the economy. Given that foreign trade is subject to international market fluctuations, which to a large extent volatize out of the scope of domestic measures, the country's exposure increases (Briguglio 2002), and small-scale economic agents turn into mere price takers without much ability to influence international prices. In this regard, the size of foreign trade of Mexico (exports plus imports) compared to GDP shifted from 43% before NAFTA implementation (1993) to 98% six years later (using data from INEGI 2000). It has stressed the Mexican agriculture as the rise of grains imports from the USA has led to lower prices.

3.4.3 Negative terms of trade

Another stressor to the countryside is the negative Terms of Trade both of the whole agricultural sector as well as of most crops within that sector. Incomes from exports depends both on the volume of these exports sold abroad and also on the price paid for them. The ratio between the price of a typical unit of exports and the price of a typical unit of imports is called commodity Terms of Trade. If the price of the country's exports are falling relatively to the price of the products it imports, it should either sell that much more of its export product (enlisting more of its scarce productive resources) or to contract debt at the amount of the net loss, merely to keep constant imported goods purchased in the past. When it succeeds, commodity terms of trade are said to deteriorate for a country. Todaro (2000: 466) conducted empirical studies about it, which suggest that relative primary-products prices have declined to manufacture goods at world level during the XX century. Between 1977 and 1994 the prices of non-oil primary products relative to those of exported manufactures declined by almost 60%. For reasons like that, countries depending on primary commodities to export must sell greater quantities in order to purchase a given quantity of imports (generally manufactures). Given that in the short-run a quick expansion of exports is usually not possible due to existing duties and other barriers to penetration of foreign markets, export amounts change only moderately, and in consequence a net loss from foreign trade appears necessarily in this kind of countries. Especially high is this risk for mono-crop countries. As Benson and Clay (2002: 23) point out, mono-crop regions must face vulnerability from not only natural disasters, but also those associated to world trade uncertainty due to non-diversified export products of primary commodities. In addition, the Prebisch-Singer thesis (Prebisch 1950; Singer 1950) argues that there was and would continue to be a secular decline in the terms of trade of primary-commodity exporters due to a combination of low income and price-elasticity of demand. This decline resulted in a long-term transfer of income from poor to rich countries, says Prebisch, which could only be combated by protecting domestic manufactures industries through the so-called process of import substitution.

In order to reduce the vulnerability associated to negative terms of trade, the Mexican economy opted for reducing the ratio of agricultural/industrial products. Roughly speaking, between 1945 and 1975 the Mexican economy applied a model of import substitution, but it was no longer sustainable due to self limits of the model as well as to world context. However, the Mexican economy has been conducting efforts towards vulnerability reduction by improving terms of trade. By 1974, Mexico joined the ranks of the new industrialized countries (NIC´s), with manufactured goods representing over 50% of total exports and agricultural goods falling back to 39% -and to 4% in 2002. In addition, whereas oil exports represented 75% of Mexico's foreign exchange earnings in 1980, by 2000 this percentage was reduced to only 18%, decreasing thus vulnerability to sudden oil prices reductions. It is true that vulnerability to negative terms of trade was reduced, but the traditional rural economy increased later its vulnerability to imports due to trade liberalization, increasing natural hazards affectation, and the dismantling of public companies aimed at supporting adaptive capacity in the agricultural sector.

3.4.4 NAFTA and agriculture

Despite surprisingly significant increases in exports due to signed trade agreements, its gains have been unequally distributed in Mexico, whereas agricultural imports (mostly grains) have contributed to drive farmers out of business. It is so partly because of *per se* asymmetries in the country, and partly because of a public planning that has ignored (or underestimated) some crucial elements of the country's vulnerability.

It might be risky to provide a judgment on the gains and losses of trade liberalization as a whole to the Mexican economy. Just 11 months after NAFTA was initiated, a dramatic economic crisis hit the Mexican economy, making it hard to disentangle the effects of the trade agreement on the economy from other factors.

The impact of trade liberalization on Mexican agriculture has received a lot of criticism even prior to the NAFTA implementation in 1994 (i.e. Calva 2004; Baffes and Meerman 1998; Levy and Van Wijnbergen 1994; Burfisher et al. 1992), as well as more positive judgments from others (see Lederman et al 2003 and Yúnez-Naude and Barceinas 2003). It has also become a very controversial point in the Mexican political arena, especially regarding the liberalization of certain sensitive products for Mexico implemented from last January 2003 due to its associated effects on small-scale farmers.

3.4.5 Unfair trade

The exposure to foreign markets volatility discussed in the previous section seems to be not as damaging as trade unfairness. Trade unfairness is driven mainly by trade partner countries, but it is influenced by domestic factors as well.

There is an increasing pressure from the Mexican poorest farmers to force developed countries to make fair trade. Fare trade is focused on cutting back on subsidies in order to reduce damage they inflict on depressing prices and, in turn, on the income of poor Mexican farmers. By 2002, over 90% of total foreign trade of Mexico was held with the United States and the European Union. Farmers from these countries have gotten more than US$ 300 billion in subsidies every year, which allows industrial-size farms to produce more hectares of crops than needed for domestic consumption, and excess production is sold overseas at prices far below the international benchmark price (Saldaña-Zorrilla 2003). Usually, excess production from rich and developed countries is big enough to drag down the world price of the commodity. On the one hand, farmers in developing countries like Mexico cannot compete with cheap imports, and on the other hand they cannot compete in foreign markets given depressed world price of the commodity. So, they lose both in domestic markets and have little chance of exporting. If one adds this problematic to the negative terms of trade of primary goods, one can realize the adverse trade conditions farmers should cope with –not only from Mexico, but also from most developing countries.

High subsidies for agricultural producers from Mexico's main trade partners make Mexican small farmers more vulnerable to imports rise. Small-farmers' discontent is growing in all developing countries. Not only Mexican farmers participated actively in the demonstrations against unfair trade at the Ministers Meeting of the World Trade Organization on September 2003 in Cancun, but also farmers from over 30 developing countries. More important, ministers of trade from most of these developing countries at the WTO have started supporting that discontent, and disagreed with moving towards a global liberalized trade in other industries (Round of Doha) as long as developed countries (centered in basically the USA, Canada, EU, and Japan) do not lift high agricultural subsidies. After the experience in Cancun, that discontent has become so generalized that forced the EU Commission, with its conditions, to negotiate with other blocks of industrialized countries an eventual subsidies reduction.

Though the external framework and international trade legislation are crucial to define foreign agricultural trade, domestic factors might also become decisive. Macroeconomic stability plays a key role in agricultural vulnerability as well. For instance, Yúnez-Naude (2001) shows that a substantial share of the variations of domestic agricultural prices during 1980-1999 in Mexico were due to rather sudden exchange-rate fluctuations.

In addition to high subsidies, there is also an increasing discontent in Mexican agricultural producers because of dumping practices, smuggling, and trade triangulation.[21] It contributes to increase imports, to depress prices, and, in last instance, to drive Mexican agricultural producers out of business, contributing to poverty and inequity in the countryside.

---

[21] Trade triangulation is considered a kind of documented smuggling, since it consists of importing merchandise from a non-NAFTA country, but presenting false purchase bills from a NAFTA area country in order to get the tariff exemption agreed in NAFTA.

3.4.6   Increasing imports, decreasing prices
in the agricultural sector

Overall, Mexican agricultural trade of balance has a deficit (see Fig. 3.6 below), which may be greater if one could integrate smuggling into the equation. After 1995, both agricultural imports and exports increased, but a slight deficit has been permanent. Most of the increase in agricultural imports is due to grains, historically produced by subsistence farmers with low yields, and much more efficiently produced in the USA at lower cost. It has led to increase imports biased to grains, with a consequent impact in reducing mean crop prices.

Price drops of agricultural grains explain over 60% of the tendency observed in Fig. 3.7 below. It is crucial given that grains like maize, beans and rice are the main crops of most subsistence farmers in this country (ECLAC 2001, 2003; World Bank 1994), undermining rural incomes of most farmers given their impossibility to increase neither their productivity nor cropping land.[22] Graph 10.a plots mean weighted agricultural prices, which are estimations of the author based on registers of mean rural prices and production with data from SIACON (2005).[23] It is made of the mean prices of crops weighted by their share production in the corresponding year at 2002 constant prices. Graph 10.b attempts at showing the net prices evolution after incorporating changes in mean yield –tons/hectare- over the same period.

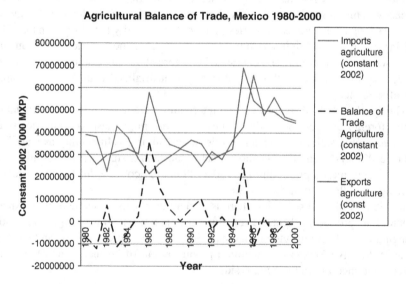

Fig. 3.6 Agricultural trade of balance, Mexico 1980-2000

---

[22] More detailed discussion in Chapter 4 of this work

[23] *Subsistema de Informacion Agricola*, Ministry of Agriculture.

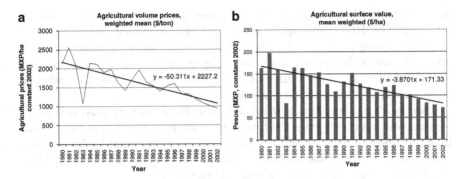

Fig. 3.7 Decreasing agricultural prices (1980-2002)

This graph thus plots weighted prices of a crop hectare in the country.[24] As observed by the slope difference of trend lines between these graphics, the price decrease in terms of production volume ($/ton) is higher than those of cropping hectares ($/ha), given the incorporation of the yield increase over this period. It means that even if one includes the yield increase, prices trend has still decreased remarkably over this period, decreasing incomes of farmers unable to crop more intensively and/or extensively, sharpening poverty conditions, and stressing particularly those farmers more frequently affected by disasters.

Finally, the economic policy reorientation implemented since the end of the 1980s has contributed to reduce farmers' adaptive capacity as it has dismantled some crucial instruments aimed at supporting agricultural productivity. One has to admit that it has helped to reduce excessive paternalism and fiscal burden, but it has contributed to reduce the adaptive capacity of farmers to reorganize in light of current competition in the said context of globalization. This discussion has been left to the next chapter, which assesses public intervention in building adaptive capacity in this country.

3.5   Conclusions

The increasing frequency of natural hazards over the past 35 years in Mexico has implied increasing economic losses as well. Hurricanes and floods have been the most damaging disaster type, mostly affecting subsistence farmers, whereas earthquakes have mainly hit the urban poor. Most natural disasters in Mexico have caused destruction of rural assets to a large extent in mainly marginalized regions. Though economic losses from extreme events have been higher in urban assets, the relative impact of disasters in marginalized rural areas is higher, contributing to increases in

---

[24] Obtained by multiplying weighted production prices ($/ton) by weighted agricultural yield (ton/ha).

rural-urban migration. Along with the higher exposure of agriculture to droughts, as the case that ca. 80% of agricultural surface in Mexico is rainfed, prevailing asymmetries in coping capacity and access to adaptive instruments in those regions is often observed, as for instance when comparing effects from hurricane Wilma and Stan in 2005 in Cancun and Chiapas, respectively, as well as among rich and poor countries sharing a same natural extreme event.

There exists *per se* a trend of the poor to settle on disaster-prone areas in urban Mexico, which more recently newcomers from the countryside have aggravated. Ironically, most of these immigrants decided to leave their rural livelihoods as a consequence of unfavorable conditions to improve agriculture-derived incomes due to a number of hazards, disaster occurrence among them.

Along with recurrent natural hazards in this country, current economic policy has contributed to stress agricultural livelihoods. The overall vulnerability of the Mexican economy has increased as the weight of foreign trade to GDP has increased as well. It has led to increasingly expose the economy to external forces. Low prices of agricultural goods in Mexico's main trade partners –most likely due to high subsidies-, has resulted in higher agricultural imports, pressing downwards agricultural domestic prices in the country. Rural incomes tend thus to decrease to those farmers without possibilities to either increase productive yield or to enlarge cropping area, like subsistence farmers. It is leading to reduce subsistence farmers' ability to create a financial pool to face hazards their activities imply, drawing a vicious circle of low income, low coping capacity, lack of climate adaptive instruments (i.e. crop insurance, reserve fund), and higher disasters vulnerability. In virtue of these facts, an assessment of the public policy and its derived intervention seem to be the next step toward reducing economic vulnerability to these hazards, presented in the following chapter. It analyzes the evolution of public intervention in improving coping and adaptive capacity through social programs and disaster management instruments, emphasizing the scope of loss sharing and risk transfer instruments.

Chapter References

Acosta C (2006) Yo, el más grande (I, the biggest one). Proceso Magazine. 30th Anniversary Special Issue, October-December 2006. Mexico City

Albalá-Bertrand JM (1993) The political economy of large natural disasters. Clarendon Press, Oxford

Baffes J, Meerman J (1998) From Prices to Incomes: Agricultural Subsidization without Protection? World Bank Research Observer 13(2)

Benson C, Clay E (2002) Vulnerability and the Global Economy. In: The Future of Disaster Risk: Building Safer Cities, Conference Papers. Kreimer A, Arnold M, Carlin A (Ed) The World Bank. Washington, DC

Bitrán D et al (2001) Características del impacto socioeconómico de los principales desastres ocurridos en México en el periodo 1980-1999. Centro Nacional de Prevención de Desastres. Secretaría de Gobernación, Mexico City

Briguglio L (2002) The Economic Vulnerability of Small Island Developing States. In: Sustainable Development for Island Societies: Taiwan and the World, Asia Pacific Research Program w/SARCS Secretariat Publication. Taiwan

Burfisher M, Robinson S, Thierfelder K (1992) Agricultural and Food Policies in a United States-Mexico Free Trade Agreement. North American Journal of Economics and Finance 3:117–139

Butterfield G (1998) Workers World. Hurricane Georges: A tale of two systems. October

Calderón E, Cazés D (1991) Prontuario de Resultados. Elecciones Federales de 1988. Fundación Arturo Rosenblueth. Mexico City

Calva JL (2004) Ajuste Estructural y TLCAN: efectos en la agricultura mexicana y reflexiones sobre el ALCA. In: El Cotidiano No. 124. Universidad Autónoma Metropolitana. Mexico City

Cannon T (1994) Vulnerability Analysis and the Explanation of 'Natural' Disasters. In: Varley, Anne. Disasters, development and environment. Ed. John Willey & Sons. Chichester

Castells M (1996) The Rise of the Network Society. Blackwell, Oxford

Castillo H (1985) Vivir el horror (Living the horror). In: Proceso No. 465, 28/sep/1985. Mexico City

CENAPRED (2001) Diagnóstico de peligros e identificación de riesgos de desastres en México: Atlas Nacional de Riesgos de la República Mexicana. Centro Nacional de Prevención de Desastres, Ministry of the Interior of Mexico, Mexico City

Center of Migration Studies of Washington (2003) Population Growth, Immigration, and the Problem of Sprawl. August 2003. By Roy Beck, Leon Kolankiewicz, and Steven A. Camarota http://www.cis.org/articles/2003/sprawlcoverage.html

CNA-Comision Nacional del Agua (1998) Reporte Anual de Actividades. Gobierno Federal Mexicano. Mexico City

Conde C, Saldaña-Zorrilla SO (2008) Climate Change in Latin-America and Caribbean Countries: Impacts, Vulnerability and the need for Adaptation. In: Challenges on Climate Change in Latin-America. The International Development Research Centre (IDRC-Canada) and Department for international Development (DFID-UK). 17 pp. Enero 2008

Currie C, Hyndman R, Wang K, Kostoglodov V (2002) Thermal models of the Mexico subduction zone: Implications for the megathrust seismogenic zone: Journal of Geophysical Research, 107 (B12), doi: 10.1029/2001JB000886, 15-1-15-13

Dussel E (2000) El Tratado de Libre Comercio de Norteamérica y el Desempeño de la Economía en México. Naciones Unidas. Comisión Económica para América Latina y el Caribe (CEPAL). Mexico City

ECLAC (2006) México: Crecimiento Agropecuario, TLCAN, Capital Humano y Gestión del Riesgo. United Nations Economic Commission for Latin America and the Caribbean, Mexico City

ECLAC (2003) Panorama Social de América Latina 2002-2003. Pobreza y distribución del ingreso. United Nations Economic Commission for Latin America and the Caribbean, Santiago de Chile

ECLAC (2001) Instituciones y pobreza rurales en México y Centroamérica. United Nations Economic Commission for Latin America and the Caribbean, Mexico City, p 30

ECLAC (1986) El terremoto de 1986 en San Salvador: daños, repercusiones y ayuda requerida. United Nations Economic Commission for Latin America and the Caribbean, , Mexico City

Frankel JA, Romer D (2000) Does Trade Cause Growth. American Economic Review 89(3):379–399

Freeman P, Mechler R (2001) Public sector risk management in Mexico for Natural Disaster Losses. Issues paper for a Wharton-World Bank Conference on "Innovations in Managing Catastrophic Risks: how can they help the poor?" Washington, DC

García V, De la Parra M (2002) Desinventar y los registros sobre escasez de agua en México. 1970-2000. In Nueva época. Año 7, septiembre-diciembre 2002. CIESA. Mexico City

Guy Carpenter (2006) Tropical Cyclone Review 2005. Instrat Briefing. January 2006 http://gcportal.guycarp.com/portal/extranet/popup/pdf/GCBriefings/Tropical_Cyclone_Review_2005.pdf

Held D, McGrew A (2000) The great globalization debate: an introduction. In: Reader TGT (ed) An Introduction to the Globalization Debate. Polity Press, Oxford, UK, pp 1–45

Helleiner E (1997) Braudellian Reflections on Economic Globalization: the historian as pioneer. In: Gill S, Mittleman J (eds) Innovation and transformation in International Studies. Cambridge University Press, Cambridge

INEGI (2006) National Institute of Statistics, Geography and Informatics http://www.inegi.gob.mx Accessed 21 September 2006

INEGI (2005) National Institute of Statistics, Geography, and Informatics http://www.inegi.gob. mx Accessed 21 May 2005

La Red (2004) Social Studies Network for Disaster Prevention in Latin-America. Panama. www. desinventar.org Accessed 15 April 2004

Lederman D et al (2003) Lessons learned from NAFTA. NAFTA's remaining trade barriers. The World Bank Group, Washington, DC

Levy S, Van Wijnbergen S (1994) Labor Markets, Migration, and Welfare: agriculture in the North American Free Trade Agreement. Journal of Development Economics 43:73–276

Micco A, Perez N (2001) Maritime transport costs and port efficiency. Inter-American Development Bank, Washington, DC

Monsiváis C (1987) Los días del temblor. In: Entrada libre: Crónicas de la sociedad que se organiza, Ed. Era, Mexico City

Munich Re (2005) Net Cat Service http://www.munichre.com/ Accessed 7 November 2005

Norris, FH, Stevens, SP, Pfefferbaum et al (2008) Community resilience as a Metaphor, Theory, Set of Capacities and Strategy for Disasters Readyness. American Journal of Community Psychology (41) 127-150

O'Brien K et al (2005) Hurricane Ktrina Reveals Challenges to Human Security. Aviso. Issue No. 14. October 2005. GECHS International Project Office. Oslo, Norway

Prebisch R (1950) The economic development of Latin America and its principal problems. United Nations, New York

Quiroz S (2002) Competitividad e Inversión Extranjera Directa en México. In: Análisis Económico. No. 37, Vol. XVIII. Universidad Autónoma Metropolitana-Azcapotzalco. Mexico City

Ramírez CJ (2005) Cuando los ciudadanos tomaron la ciudad en sus manos. In: La Jornada, News. September 11, 2005. Mexico City

Robinson S, Burfisher M, Thierfelder K (1995) The impact of the Mexican crisis on trade, agriculture, and migration. Trade and Macroeconomic Division, International Food Policy Research Institute, Washington, DC

Saldaña-Zorrilla SO (2003) El Comercio México-Unión Europea. In: Análisis Económico No. 37., Vol. XVIII. 1er. Semestre 2003. Universidad Autónoma Metropolitana-Azcapotzalco. Mexico City

SIACON (2005) Sistema de Información Agropecuaria de Consulta. Ministry of Agriculture and Livestock http://www.siap.sagarpa.gob.mx/sistemas/siacon/SIACON.html Accessed 5 May 2005

Singer H (1950) The distribution of gains between borrowing and investing countries. American Economic Review 40:473–485

Singh MP, Suárez LE (1987) Seismic response analysis of structure-equipment systems with neoclassical damping effects. Earthquake Eng Structure Dyn 15:871–888

Todaro MP (2000) Urbanization and Rural-Urban Migration: Theory and Practice. In: Economic Development. Seventh Edition. Addison Wesley Longman. Essex

US Census Bureau (2002) http://www.census.gov/acs/www/ Accessed 2 August 2002

Vatsa KS, Krimgold F (2000) Financing Disaster Mitigation for the Poor. In: Managing Disaster Risk in Emerging Economies. The World Bank

World Bank (2002) Country assistance strategy of the World Bank Group for the Mexican United States, 2002. Washington, DC

World Bank (1994) Staff Appraisal Report of Rainfed Areas Development Project June 1994. Agriculture Operation Division. Country Department II. Latin-America and the Caribbean Regional Office

World Bank, Guy Carpenter, IIASA (2000) Managing the financial impacts of natural disaster losses in Mexico. Report prepared for the Government of Mexico

Yúnez-Naude A (2001) Mexican Agricultural Change and Mexico-US Inmigration. Issues paper presented at the Conference of the Project Changing Face of Rural California, University of California, Davis, Holtville, CA. Zambia and Zimbabwe, ed. Valdes, A. and Muir-Leresche, K., Washington, DC

Yúnez-Naude A, Barceinas F (2003) Los Impactos del TLCAN (NAFTA) en el sector agrícola de México. Mimeographed. Department of Economics. El Colegio de México, México City

Chapter 4
Disaster risk management assessment

Wenn wir in der Lage kommen, unsere Grundrechte verteidigen
zu müssen, so 'dürfen' wir weder Feindseligkeit noch
Gleichgültigkeit vom Staat erfahren, sondern Wohlwollen!

KARL POPPER, Ministaat oder paternalistischer Staat?

Abstract Despite its novel design, the still low penetration of governmental instruments for disaster risk reduction in Mexico has led to society's high reliance on post-disaster measures. It has encouraged moral hazard among potential victims. Even though crop insurance, the most implemented ex-ante instrument, has increased coverage over the past decade, public disaster instruments for risk reduction are still being underemployed. Accessing public risk reduction funding requires project proposals from national and sub-national governments based on concrete risk assessments. However, the prevailing lack of proposals from provincial and municipal authorities explain such low implementation. In this regard, this chapter provides a set of suggestions aimed at integrating risk management, increasing risk reduction measures and planning in a more –even if centralized- integral manner.

Keywords natural hazards • disasters • weather • climate • financial instruments • insurance • economic planning and development • risk transfer • loss sharing

4.1 Introduction

Mexico is a country highly exposed to a wide variety of natural hazards. The southern Pacific coast of the country belongs to the so-called Ring of Fire of tectonic activity. Tropical storms and hurricanes are a constant hazard on both the Pacific and Atlantic coasts. Droughts increasingly afflect the whole country, even with rising presence in typically wet-regions, and in addition a desertification process is hitting vast regions of the north. As such, natural hazards in Mexico have increased their economic impact over the past four decades, as analyzed in Chapter 3.

© The Author 2015
S.O. Saldaña-Zorrilla, *Natural Disasters, Foreign Trade and Agriculture in Mexico,*
SpringerBriefs in Environmental Science, DOI 10.1007/978-3-319-17359-7_4

This chapter explores current actions that the government implements to manage disaster risk of natural hazards in Mexico. A recent OECD (2013) review of the Mexican National Civil Protection System points out the need to change their approach to outweigh risk prevention and climate change adaptation in developing a strategic vision for the country. This is particularly interesting if we consider the increase in the severity of extreme weather events as a result of climate change predicted by most of the global scientific community (ECLAC 2010; IPCC 2007; Stern 2007); they warn of the social cost that will continue to rise in the absence of efforts to reduce the risk and vulnerability of the population.

There is, however, no recent integral assessment of disaster risk in Mexico. Although the above mentioned OECD review (2013) of the National Civil Protection System provides an interesting assessment, its analysis leaves out catastrophic agricultural insurance, a critical part of comprehensive risk management of a country. On the other hand, there are recent evaluations of public programs for disaster risk management (COLMEX 2013; CONEVAL 2010, 2011, 2013; UACH 2005, 2006, 2007, 2010), but these consist of only individual program evaluations, lacking integrative and comparative analysis. Thus, this chapter aims to give a comprehensive view of government risk management to conclude with a series of policy recommendations.

4.2 Definition of terms and summary of public instruments for disaster risk management

The Mexican government's intervention in the field of disasters goes beyond civil protection, as it includes a complete variety of public instruments for disaster risk management. A public instrument is a legal term that is used for any formally executed written document that can be attributed to the public sector that formally expresses a legally enforceable act, process or contractual duty, obligation, or right (Grifis 2010). Risk management is a set of pre- and post-disaster actions towards facing negative consequences of hazards (Freeman and Mechler 2001). Pre-disaster actions include risk transfer (i.e. insurance), risk assessment (monitoring, mapping, etc.), mitigation works (i.e. infrastructure strengthening, etc.), and preparedness. Post-disaster actions include emergency response (aid, clean-up, etc) and loss sharing for rehabilitation and reconstruction (i.e. private-public and national-local resources). For its part, civil protection is a concept embracing both pre- and post-disaster. Derived from the civil defense terminology, civil protection is the network of awareness and emergency response for disaster and crisis state avoidance, initially based on various technical interventions usually known as preparedness. Preparedness is the management capability before a disaster occurs to provide an effective and efficient (prompt and due) reaction to face it. (Freeman et al. 2002).

Disaster risk reduction is the outcome of the corresponding instruments' actions. In terms of impact, depending on their efficacy, these said public instruments for disaster risk management may reduce, maintain or even increase risk. *Risk* can be expressed as the probability of an undesired outcome, based on the potential occurrence of harmful events and on susceptibility to them among those likely to be exposed (Dilley and Boudreau 2001). *Risk reduction* is, thus, the outcome of decreasing the said probability through actions from a certain set of instruments.

In Mexico, monitoring, early warning and immediate relief are made by the National Civil Protection System (SINAPROC, for its acronym in Spanish) through its various subsidiary entities as the National Center for Disaster Prevention (CENAPRED) and the civil protection systems of the states and municipalities. Along with SINAPROC, the National Defense Ministry (SEDENA) provides immediate relief as well through the so-called National Defense Plan III (Plan DN III); for its part, loss sharing is mainly funded through the federal Fund for Disasters, FONDEN, supplemented with a combination of measures, such as budget diversion, catastrophic bonds (FONDEN XL) and reinsurance; the disaster mitigation works can be funded with a federation-provinces-municipalities mix fund known as FOPREDEN; the waterworks for reducing flood and drought risks are undertaken by the National Water Commission, CONAGUA, in coordination with state and local governments; the risk transfer is done through grants for hiring agricultural-parametric insurance premiums through AGROASEMEX, the public insurer for the overall countryside, while insuring in the low-income agricultural sector is made by the Ministry of Agriculture's (SAGARPA) CADENA program. These instruments are summarized in Table 4.1.

While the Mexican government has achieved remarkable progress in the implementation of monitoring and early warning systems as well as emergency response (OECD 2013; Saldaña-Zorrilla 2007; World Bank 2000), the pre-disaster instruments for risk management still present enormous challenges. This chapter warns of the excessive budget allocation on *ex-post* instruments and the consequent

Table 4.1 Summary of public instruments for disaster risk management (2013)

	INSTRUMENT			
	Pre-disaster			*Post-disaster*
OBJECTIVE	Risk assessment and preparedness	Mitigation	Risk transfer	Emergency response and loss sharing (Reconstruction and rehabilitation)
Private assets	SINAPROC	- FOPREDEN - CONAGUA - PIARSE	- AGROASEMEX - CADENA	CADENA
Public assets	SINAPROC	- FOPREDEN - CONAGUA	- Private insurance - Reinsurance FONDEN-XL	FONDEN
Immediate aid	SINAPROC	N.A.	N.A.	-SEDENA -SINAPROC - FONDEN - CADENA

SOURCE: Own elaboration.
N.A. = Not applicable.

under-utilization of ex-ante instruments. As seen in further sections of this book, implementation of disaster risk reduction works is scattered in different areas of government without a strategic plan. Reconstruction and rehabilitation spending is comparatively high: for every dollar invested in disaster risk reduction, 37 dollars is spent in reconstruction, with a number of drawbacks as discussed below. There is a disconnect between the agricultural insurance coverage and risk reduction public works as well, so that, overall, this chapter emphasizes the need to integrate risk management instruments to thereby prevent transients from serving as mere instruments: they should work together in reducing risk on a measurable basis.

It suggests, among others: i) Investing in areas where public investment can render greater effects for capacity building for response and adaptation to natural hazards, especially among the most vulnerable population and economic activities; ii) Regarding disaster management, this chapter presents a discussion on some negative results of the still unfinished administrative decentralization process in Mexico of the last two decades.

Section 4.3 deals with a loss-sharing scheme, briefly exposing the features and performance of FONDEN. Section 4.4 addresses public instruments of risk transfer, explaining the evolution through the last four decades of AGROASEMEX as well as the structure of the CADENA program and its predecessors (PACC and FAPRACC). Section 4.5 describes the operation of FOPREDEN's, a disaster risk reduction instrument. Section 4.6 concludes with a wrap on the complementary nature of the analyzed instruments, pointing as well to the various levels of government on areas where further efforts are needed.

4.3   Loss sharing

4.3.1   FONDEN

The National Disaster Fund (FONDEN, for its acronym in Spanish) is a federal government fund to enable rapid access to financial resources for disaster recovery by various government agencies at all three levels of government, covering exclusively public assets. This fund aims to give priority to the reconstruction of public assets involving poor families. It is managed by the Ministry of the Interior (SEGOB, for its acronym in Spanish) in close coordination with the Ministry of Finance and Public Credit (SHCP, for its acronym in Spanish).

4.3.1.1   CatMex catastrophe bonds and MultiCat

Mexico pioneered the first issue of a parametric instrument in a Latin American country: a catastrophe bond, the CatMex bond. Created on May 2006, this bond triggers a payment to the bond holders from the FONDEN if the following two conditions are met: i) the issuance of an official declaration of a state of disaster emergency by the Ministry of the Interior and ii) the registration of an earthquake, magnitude or epicenter in any of the specified areas. The parametric trigger has not been activated yet. In 2010, this catastrophe bond was issued again, expanding its coverage to embrace earthquakes and hurricanes, issuing compensation worth 290 million U.S. dollars. In 2012, Mexico launched MultiCat, another catastrophic bond contract for three more years with expanded coverage of earthquake risk in most areas of the country and covering the risk of hurricanes and tropical storms in specific coastal areas of the Pacific and the Atlantic Ocean. Covered value reaches 315 million U.S. dollars, of which, 140 for earthquake, 100 million for tropical storms in the Pacific coast and, the remaining 75 million, for Atlantic hurricanes. The cumulative premium paid during the period 2009-2012 was equivalent to 95 million (SHCP 2013; OECD 2013).

4.3.1.2   FONDEN-XL Reinsurance

From 2011 until 2013, FONDEN arranged a reinsurance scheme known as excess loss (XL) to cover disaster losses exceeding one million dollars. This scheme covers a surplus of 400 million U.S. dollars (ca. 4,800 million Mexican pesos: SHCP 2013), with an annual premium of US$ 100 million. It represents 25% of coverage. Although this high premium value seems to be too high, however this is not entirely far-fetched once considered that in 2010, 2007, 2005 and 1999 the value of FONDEN payments exceeded that threshold. Facing such a high recurrent disaster losses warns of the need for reducing the risk in a structural way, since risk transfer –due to the very likely probability of high losses- is very expensive and does not reduce at all the level of risk of these assets into the future.

4.3.1.3   Challenges

As part of a broader decentralization process increasingly implemented over the past decade in Mexico, FONDEN has been decentralizing duties to provincial and municipality level. In order to go further compromising province and federal agencies in managing their own risk, it is now mandatory for accessing FONDEN from province governments the hiring of insurance premiums for its own assets. However, the federal government continues to assume part of the risk in the repair and reconstruction of the socially strategic local infrastructure owned by sub-national governments even if not fulfilling the said requirement. In accordance with FONDEN Rules of Operation, the federation, provinces and municipalities can share some responsibilities, such as the reconstruction or repair of transportation infrastructure. However, due to high debt and tax collection shortages in many provinces, it is still a common practice that the federation assumes almost the entire costs in these areas.

4.4   Risk transfer

Risk-transfer from natural hazards in Mexico is concentrated almost entirely in the agricultural insurance market. 64% of agricultural insurance premiums are subsidized by the government (CONEVAL 2011) and the national agricultural area covered by any insurance scheme has grown significantly over the last ten years. The main scheme of agricultural insurance subsidy is managed by AGROASEMEX and the CADENA program, which are set forth below to further take stock of its achievements and challenges.

4.4.1   AGROASEMEX

AGROASEMEX is the government company responsible for managing the "Subsidy Program of Agricultural Insurance Premium" (PSPSA, for its acronym in Spanish), which contributes to the establishment of a national system of risk management and the expansion of agricultural insurance. The fulfillment of the objective of the program is done through: (i) supporting farmers in reducing the cost of premiums paid in hiring agricultural insurance, (ii) encouraging the participation of private and social sectors in the agricultural insurance, and (iii) promoting the development and strengthening of prevention and protection mechanisms for agricultural producers nationwide to cope with the risks faced by their activity (DOF 2013b). AGROASEMEX is responsible for advising the parametric insurance scheme of SAGARPA as well –analyzed more in detail below- and serves as reinsurer for private insurers. AGROASEMEX uses a scheme known as *Fondos*, consisting of mutuality associations of farmers and/or ranchers organized as for the establishment and implementation of micro-insurance (DOF 2013b). The risks covered by the *Fondos* include drought, excessive moisture (due to floods, heavy rains, etc.), frost, hail, fire, high winds, pests, failure to grow, lack of germination, death of livestock, accident, disability and forced slaughter of animals. Multi-peril coverage of *Fondos* includes biological disasters and economic dangers as well –such as increases in production costs due to increases in the price of inputs and risks related to sales revenue. The federal budget allocated to AGROASEMEX has been increasing. Between 2005 and 2011 it had a cumulative increase of 48% in real terms, with a real growth rate of 8% annually (Figure 4.1).

Similarly, the agricultural subsidy coverage has been increasing. In 2013, the agricultural area that received some type of government subsidy for hiring insurance premiums amounted to 12 million hectares, contrasting with the 3 million hectares covered at the beginning of the last decade. In 2010, coverage reached 255,464 producers in 1,360 municipalities throughout the 32 provinces of the country. In the same year, 63% of agricultural insurance premiums had government subsidies (CONEVAL 2011). Meanwhile, the fiscal cost per hectare was strongly depressed, going from 1,622 pesos (constant 2013) in the early 1980's to 540 in the 1990s and 315 in 2009 (the last reported period) (Figure 4.2).

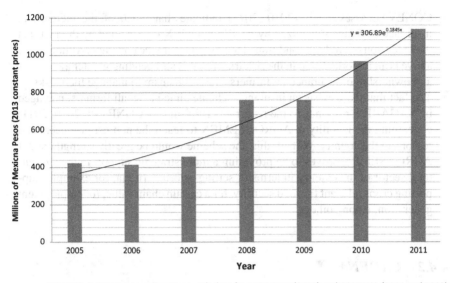

SOURCE: Author's own estimations with data from COLMEX (2012) and CONEVAL (2011 and 2010).

Fig. 4.1 AGROASEMEX Budget 2005-2010

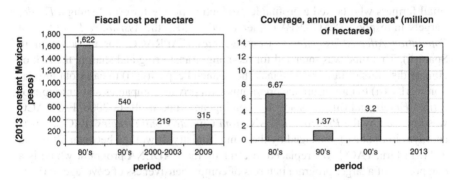

SOURCES: Author's own estimation with data from INEGI, CONEVAL and AGROASEMEX.
*The 1980's correspond to ANAGSA, 90's to AGROASEMEX and 00's and 2013 to the mix of
AGROASEMEX (*Fondos* and direct subsidies) and SAGARPA (FAPRACC, PACC and CADENA).

Fig. 4.2 Crop insurance subsidy coverage

4.4.1.1   Challenges

The main challenges of AGROSASEMEX are summarized in the following points:

- There is an absence of studies for determining the causes of the, besides AGROASEMEX, low level of agricultural insurance in Mexico. Literature in this regard just argues it is due to the prevailing high premium prices, which in turn is due to the latent high level of risk (COLMEX, 2013).

- CONEVAL assessments (2010, 2011) indicate that much of the increase in insurance coverage subsidized by AGROASEMEX is explained by the fact that the extension of credit is conditioned to having insurance coverage. Although the above is an important step in the risk transfer, it is still pending greater awareness and ownership of insurance by farmers as a mechanism to manage their risks.
- The fact that there exists an insurance scheme aimed at small farmers, managed by SAGARPA, responds to the recurrent criticism (i.e. CONEVAL 2010, 2011) concerning the inability of AGROASEMEX to reach the poorest producers and regions in the country. With all this, the scheme would be greatly strengthened if AGROASEMEX proves to be providing a more strategic support to the agricultural sector of the country in terms of setting priority criteria, for example, supporting crops relevant for food security, for a certain chain value, for employment generation, among others.

4.4.2   CADENA

One of the main remarks concerning AGROASEMEX's performance during the 1990's and the early 2000's points out that even when *Fondos* allowed the development of a more efficient agricultural insurance market, on the other hand it left out small farmers who lacked a profitable production profile for establishing a *Fondo*. Largely in response, SAGARPA implemented in 2003 the *Assistance Fund for the Population Affected by Climate Contingencies* (FAPRACC, for its acronym in Spanish). This fund was intended for reducing, transferring and sharing the risk of climate disasters of farmers who lack the financial capacity: (i) to carry out mitigation works, (ii) to hire insurance premiums with private companies and/or to join or form a *Fondo*, and (iii) to establish a contingency fund itself. In 2008, FAPRACC was replaced by the *Program to Assist from Climate Contingencies* (PACC, for its acronym in Spanish), managed by the same Ministry. It worked between 2008 and 2011. In turn, PACC was replaced in 2011 by the CADENA program, which is a component of a larger program in terms of comprehensiveness of coverage of risks: the program for Prevention and Risk Management (PPMR).

4.4.2.1   Coverage

CADENA aims at supporting rural sector producers to relevant natural hazard damages in farming, aquaculture and fisheries. This program's support consists of post-disaster compensation in cash or check payable to eligible concerned producers, located in municipalities with recurrent natural hazards and with the partnership of provincial governments. To date, the official data unavailability concerning this program hinders for evaluating its performance. In this regard, there exists data from only the National Council for Evaluation of Social Development Policy (CONEVAL, for its acronym in Spanish) (CONEVAL 2013), which is the federal agency responsible for evaluating the various social policy programs. In this regard, CONEVAL notes that this program has flaws related to the quantification of their potential and target population as well as of socioeconomic data collected of its beneficiaries (CONEVAL 2013, p. 2). The CADENA program has continued to operate in a similar way as it had done the PACC and FAPRACC. As a result, the data produced by the latter two through their external evaluations (UACH 2010; Colegio de Posgraduados 2008; UACH 2007; UACH 2006; UACH 2005; UNAM-CCA 2004) allow for some comparisons over the past decade. Overall, FAPRACC and PACC have been relevant for post-disaster business continuity and recovery among affected population. External Assessments to PACC 2008, 2009 and 2010 provide relevant findings on this regard:

- Between 2008 and 2010 there was an increase in the relative utility of post-disaster support for beneficiaries. While the 2008 evaluation found that, after a climatologic contingency, 80% of surveyed beneficiaries stated they kept more than 25% of their productive assets working due to the governmental support, the 2009 evaluation showed that this percentage rose to 92% and to 98% in 2010 (Colegio de Posgraduados 2008) (Figure 4.3).
- The 2009 assessment revealed that, after natural hazards, 96% of the beneficiaries returned 75% of their productive assets, while this percentage is only 82% among non- beneficiaries (UACH 2010), that is, the percentage of population productively reinstated was 15% higher among beneficiaries.

From its part, the coverage of catastrophic agricultural insurance FAPRACC-PACC increased considerably as insured crop area increased from 0.11 million hectares in 2003 to 8.4 million in 2010. Meanwhile, the number of livestock units grew from 0.3 million to 4.2 million between 2006 and 2010 (UACH 2010) (Figure 4.4).

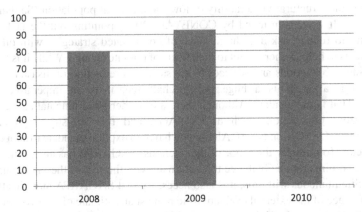

SOURCE: Author's own estimations with data from CONEVAL (2013, 2011, 2010) and Colegio de Posgraduados (2008).

Fig. 4.3 Percentage of post-disaster beneficiaries with over 25% of assets working due to PACC

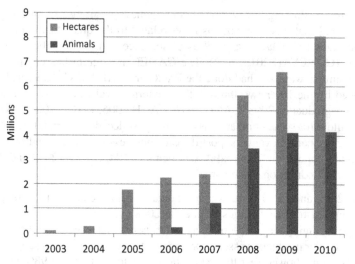

SOURCE: Author's own estimations with data from UACH 2010.

Fig. 4.4 Catastrophic insurance coverage in Mexico (2003-2010)

4.4.2.2 Challenges

On the one hand, we should recognize the recent FAPRACC-PACC dramatic increase in terms of coverage along with its relative utility of the program for its post-disaster continuation of activities among affected population. On the other hand, since its inception in 2003, FAPRACC planning has lacked a structural connection with other crucial parts of the Mexican public administration. It prevented this instrument from reducing risk and structural vulnerability of low-income rural population (Saldaña-Zorrilla 2004). It was again noted by CONEVAL (2013) pointing out the lack of infrastructure to reduce risk and the absence of coordinated strategies with other public agencies to help farmers adapt to the effects of climate change. While it is true that providing public goods and services to the structural reduction of disaster vulnerability in rural areas is a huge policy challenge, it is an imperative of SAGARPA. That is its mandate. Although there are further governmental agencies with concurrent jurisdiction -such as CONAGA and the Ministry of Social Development, SEDESOL-, it is SAGARPA which has an explicit mandate to support the rural sector. To multiply the scope of their actions, SAGARPA has an enormous political and administrative challenge to coordinate with the rest of the intervening ministries. Simultaneous with this, it is also necessary that SAGARPA modifies the terms of reference of the external evaluations to request an estimate of the evolution, impact, effectiveness and efficiency of its PPMR (and in particular the CADENA program) in reducing vulnerability faced by its potential and target population, ideally including climate variables also. This involves the application of cost-benefit analysis in order to scale up to find microeconomic impacts of natural hazards on income, output and prices: comparing a no-PPMR baseline scenario with a PPMR one: it will certainly allow for appropriate adjustments to optimize public spending.

4.4.3 Catastrophic insurance system balance

Apart from the remarks of the previous sections, the Federal Superior Audit (ASF, for its acronym in Spanish) published an analysis of government subsidies to agricultural insurance in Mexico (ASF 2010). It recognizes the utility of AGROASEMEX's PSPSA as well as the corresponding SAGARPA programs in increasing rural penetration of insurance, helping reduce the cost of insurance contracts by farmers and for post-disaster business continuity among low-income rural producers. However, ASF also made remarks about: i) the lack of control of the list of beneficiaries; ii) the absence of means of verifying compliance with the requirements of the beneficiaries, as well as; iii) lack of impact analysis on beneficiaries (ASF 2010). The insurance system reforms in the rural sector have had a positive effect which allowed further development of a private agricultural insurance market while reducing much of the past state paternalism. However, the granting of the subsidy has not, neither before nor after the reforms, been accompanied by a structural strategy for disaster prevention such as to reduce the risk associated to premium prices. Even with the lowering of the premiums price of raw competition that *Fondos* may have triggered, the absence of a strategy to reduce physical risk contributes to keeping premium prices still high. This is strengthened by the moral hazard arising from the producers' expectations derived from past post-disaster government actions (reconstruction, compensation, indemnities, etc.). It has acted as a precedent for further reliance on governmental response preventing farmers from implementing disaster prevention by them.

4.5 Risk reduction

The nature of risk reduction instruments is much more complex. Unlike risk sharing and risk transfer, risk reduction instruments rely on edge cutting investments. Risk reduction investments include both construction of new infrastructure or repair/rebuild existing infrastructure. Now, let us briefly describe the corresponding public instrument for disaster risk reduction in Mexico, FOPREDEN.

4.5.1 FOPREDEN

The Disaster Prevention Fund (FOPREDEN) is a federal fund for mitigation projects and disaster risk research. This fund is an instrument often underemployed. FOPREDEN's granting of resources based on an efficiency analysis. As stated in the last publicly available FOPREDEN's external assessment (Puente 2005), although in 2005 the budget of FOPREDEN exceeded 400 million of Mexican pesos, only a quarter of this amount was spent on projects. During the period 2004-2007, this fund had a cumulative underuse of around 259 million pesos. It equals 35% of its overall budget allocation for that period (see Chart below). Even when the unused share of the budget is transferred to the subsequent fiscal year (revolving

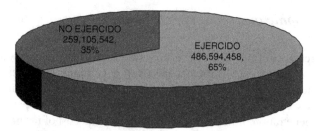

SOURCE: Author's own estimations, based on data from SEGOB
(2006), SHCP (2007, 2008) and Puente (2005).

Fig. 4.5 FOPREDEN's Cumulative Budget

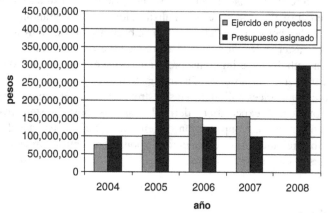

SOURCE : Author's own estimations based on data from the Ministry
of the Interior (SEGOB, 2006) and Federal Expenditure Budget
(SHCP 2008, 2007).

Fig. 4.6 Annual evolution FOPREDEN budget

fund), the remarkable fact is that these resources are not being invested in actions
and at the pace for which the fund was designed for (Figure 4.5).

This highlights the still insufficient disaster prevention projects proposals from
sub-national governments and federal agencies. Partly, it is so due to the lack of
institutional capacity of many of them, especially those small municipalities with
fewer resources.. Ironically, the said lack of institutional capacity prevails among
the most disaster prone regions. In other cases, this is due to lack of interest from
authorities concerning disaster risk. Despite the large share of its population living
in high disaster risk areas, prevention rests on a priority level so low that their pro-
fessional and material resources already fail to plan works on the subject (Saldaña-
Zorrilla 2007) (Figure 4.6).

FOPREDEN's budget for the year 2008 amounted to 300 million pesos (see figure above), but if underuse prevails, it may raise arguments to divert them to further areas of spending items. Between 2008 and 2013, FOPREDEN's annual budget remained at 300 million pesos. Since its inception, FOPREDEN has funded 130 disaster prevention projects for a cumulative total of 1,400 million pesos. More than half of these resources have been allocated to projects in the states of Tamaulipas, Tabasco, Sonora and Chiapas (OECD 2013).

4.5.1.1 Challenges

FOPREDEN still has some hurdles to overcome:

– The current FOPREDEN's scheme should move from granting resources on a kind of (rather political) on-demand basis to a risk analysis based decision making. Institutional capacity then plays a major role. However, as a result, resources allocation is not necessarily done where it is most needed, but where there have been more requests. Both the OECD report (2013) and Saldaña-Zorrilla (2007) criticize the criterion of project formulation from provincial governments as a requirement to carry on prevention measures in Mexico (according to FOPREDEN Operation Rules: DOF 2010), when it might be better to implement instead a needs-based criteria, i.e. prevailing level of risk and vulnerability.
– Concerning budget underuse, the discussion should not focus on reducing the allocation of projects to those not so affected states, but how to increase it in the most affected states to a value that keeps more proportion to its historical level of losses. With all this, the amount exercised by FOPREDEN seems to be still relatively low.
– Roughly, for every dollar that the government invests in disaster prevention, it spends 37 dollars in reconstruction. During the period 2005-2011, FOPREDEN's cumulative budget reached 140 million U.S. dollars, which compared to the FONDEN accumulated budget during the same period, U.S. $ 5,200 million (OECD 2013), equals a 1 to 37 ratio (Figure 4.7).

Fig. 4.7 Public budget distribution for disaster prevention, reconstruction and risk transfer: cumulative 2005-2011

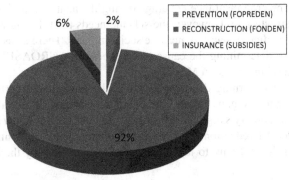

SOURCE: Based on data from OECD (2013) and AGROASEMEX (2013).

This proportion, however, might change if we include CONAGUA's, SEDESOL's and SAGARPA's projects, which, although more indirectly, implement public work for disaster risk reduction as well. Mitigation is many times included into infrastructure works, and as that, its tasks rely increasingly on local governments (state and municipal). It occurs with disaster risk mitigation works derived from current disaster prevention instruments (mainly FOPREDEN) as well, where the initiative has been left to state authorities upon request from the municipalities. Estimating the contribution of these programs for reducing disaster risk involves a detailed project-by-project analysis, which is beyond the scope of the present document.

4.6   Conclusions

As seen from the above discussion, if the herein described instruments might succeed in working properly, they might produce multiplying effects. The feasibility of investing in disaster mitigation works, insurance, contingent funds, or any other instrument, must depend on a comprehensive project development (rural-urban, public-private entities, national-province and prevention-reaction). It seems to have carried on still insufficient investments to mitigate disaster risk.

4.6.1   Government performance

Concerning loss sharing, the federal government performance has been inadequate in that of inhibiting provincial governments from undertaking self-responsibility for disaster risk reduction of their own assets by their own. The federal government continues to assume an important part of the risk in the repair and reconstruction of the socially strategic local infrastructure owned by sub-national governments even if not fulfilling the corresponding law requirements.

For its part, government performance on risk transfer has been, overall, positive. The crop insurance coverage during the analyzed period means an unprecedented growth rate. However, the scheme needs to still achieve a better control of target beneficiaries and to measure social impacts. There is, as well, an absence of studies for determining the causes of the, besides AGROASEMEX, low level of agricultural insurance in Mexico.

Concerning disaster reduction, the government performance has been insufficient. Mitigation works are being increasingly decentralized. However, relative to the country's disaster exposure, there have been still few project proposals for disaster risk reduction from provincial and municipal governments. This lack of projects is due not only to lack of interest in the subject by the local authorities, but more often it is due to the lack of institutional capacity (technical and administrative) to identify risk and propose concrete measures. This, ironically, happens predominantly in the most vulnerable municipalities with greater exposure to risk. Thus, there is a need to provide *subsidiarity*[1] to those municipalities without enough institutional capacity through federal agencies. *Subsidiarity* is an organizing principle of decentralization, stating that a matter ought to be handled by the smallest, lowest, or least centralised authority capable of addressing that matter effectively but, at the same time, the central authority should have a subsidiary function, supporting those tasks which cannot be performed effectively at a more immediate or local level.

4.6.2 Policy challenges

The following points summarize the main policy challenges for more effective risk reduction and vulnerability to disasters in Mexico:

- In general, there is a deficit of central planning from the Mexican public sector to manage disaster risks. This is evidenced by the lack of an integral assessment of the country's disaster management in the published literature, as well as by the observed insufficient integration of actions for disaster management.
- The decentralization process has transferred responsibilities to local governments too quickly, commonly exceeding its technical and administrative capacity. This has led to underutilization of existing programs for disaster prevention.
- The CADENA program devotes most of its resources to indemnities and insurance subsidies for small farmers, but its spending has a tendency to increase due to increasing climatic uncertainty so massively that its spending will be most likely difficult to control in the near future.
- As the economic impact from natural hazards in Mexico tend to dramatically increase –especially the weather-related ones-, FONDEN's budget will have to consequently continue growing unless the due disaster risk reduction measures take place in a more proactive manner from now on.
- The disaster-prone regions also tend to be less prepared for prevention and very often ignore the existence of governmental instruments in the field. These regions have poorly trained personnel to risk identification and management of resources for prevention.

---

[1] For further details: Tocqueville, Alexis de (2000). *Democracy in America*. Chicago: The University of Chicago Press. ISBN: 0-226-80532-8.

In this regard, the following suggestions might contribute to solve the above challenges at the three levels of government:

4.6.3 Federal Government

a) To link SINAPROC effectively with the corresponding agencies responsible for managing agricultural risks, as they can benefit from sharing their best practices as well as mainstreaming efforts in a more effective and strategical manner.
b) Develop a common methodology for various government agencies to identify socio-economic vulnerability, also applicable to the respective units of provincial and local governments.
c) Provide incentives to states and municipalities for the development of individual projects for disaster prevention, setting mandatory deadlines for the cases of the municipalities identified as most vulnerable, applying the principle of *subsidiarity*, this is:

    i) If these municipalities do not respond within the deadline, the province should assume that obligation in a subsidiary manner, charging the project proposal formulation costs to the corresponding municipality as a kind of financial penalty.

    ii) If either the provincial government succeeds in assuming the corresponding project proposal elaboration, the central government should take that responsibility, charging the project proposal formulation costs to the corresponding province as a kind of financial penalty. The aim of this procedure is to prevent inaction in carrying on structural projects for disaster risk reduction the country needs most.

4.6.4 Provincial Governments

a) To develop cross-section matrices of socioeconomic vulnerability and climate risks. It will allow reaching the country's most vulnerable population to natural hazards.
b) To give priority to disaster prevention projects which simultaneously trigger the local economy; it requires implementing a consistent methodology, based on a cost-benefit analysis.
c) To schedule a productive reorientation to adapt to current and most likely forthcoming climate conditions.

Municipal and local leaders:

a) In coordination with provincial governments, to train staff on disaster risk identification (on issues such as civil engineering, hydraulics, finance, etc.) and to learn how to access federal resources for disaster risk reduction via proper scientific formulation and project management. Such staff should be as permanent as possible in order to maintain continuity over administration changes.

b) To consult with the community to capture risk perception, identify legitimate needs and find participatory and realistic solutions.

c) To develop detailed risk atlas for proper risk monitoring and a precise formulation and evaluation. This implies a greater exchange of information with other government agencies and the scientific community for better database update and accuracy of methods.

4.6.5   The three levels of government

They should reduce risk simultaneously. Structural reductions of risk such as asset relocation from prone to less prone disaster areas combined with better land use, with an optimal crop programming according to the current and the most likely forthcoming weather conditions, the construction of water infrastructure (both gray and green) to reduce flooding and drought, among others, produce a direct impact on reducing disaster risk and thus put downward pressure on the cost of the insurance and reinsurance premiums for the country. It is therefore also essential to keep updated the risk atlas in order to internalize these risk reductions to upgrade the country's risk position in the upcoming premiums negotiations for contracting further catastrophic insurance and reinsurance policies.

Chapter References

ASF-Auditoría Superior de la Federación (2010) Protección al patrimonio del sector rural (Rural sector assets protection), working paper, Cuenta Pública 2009, Mexican Federal Congress, Mexico City, 15 June

Colegio de Posgraduados (2008) Evaluación externa 2008 al programa de atención a contingencias climatológicas (2008 program for climate hazards response extern assessment), working paper, Coord. Figueroa, Oscar, Colegio de Posgraduados, Mexico City, 21 May

COLMEX (2013) Evaluación de consistencia y resultados del programa de subsidio a la prima del seguro agropecuario (Assessment of consisntency and results to the crop insurance premium subsidy program), working paper, Centro de Estudios Económicos, El Colegio de México, Mexico City, 11 October

CONEVAL (2013) Evaluación de consistencia y resultados 2011-2012: programa de prevención y manejo de riesgos (2011-2012 assessment of consisntency and results: prevention and risk management program), working paper, Consejo Nacional de Evaluación de la Política de Desarrollo Social, Mexico City, 12 March

CONEVAL (2011) Informe de evaluación específica de desempeño 2010-2011 (2010-2011 specific assessment report), working paper, Consejo Nacional de Evaluación de la Política de Desarrollo Social, Mexican Federal Government, Mexico City, 28 February

CONEVAL (2010) Informe de evaluación específica de desempeño 2009-2010 (2009-2010 specific assessment report), working paper, Consejo Nacional de Evaluación de la Política de Desarrollo Social, Mexican Federal Government, Mexico City, 22 March

Dilley M, Boudreau TE (2001) "Coming to terms with vulnerability: a critique of the food security definition". Food Policy 26(3):229–247

DOF (2013a) Acuerdo por el que se dan a conocer las reglas de operación de los programas de la secretaría de agricultura, ganadería, desarrollo rural, pesca y alimentación (Agreement for promoting the operation rules from the ministry of agriculture, livestock, rural development, fisheries and food), Art. 30-36, Mexican Federal Congress, Act of Congress, Diario Oficial de la Federación, Mexico City, 11 February

DOF (2013b) Reglas de Operación de los Programas del Subsidio a la Prima del Seguro Agropecuario y de Apoyo a los Fondos de Aseguramiento Agropecuario (Operation Rules of the Crop Insurance Premium Subsidy Programs and for Supporting Crop Insurance Funds), Mexican Federal Congress, Act of Congress, Diario Oficial de la Federación, Mexico City, 8 February

DOF (2010) Reglas de Operación del Fondo para la Prevención de Desastres (Disaster Prevention Fund Operation Rules), Mexican Federal Congress, Act of Congress, Diario Oficial de la Federación, Mexico City, 23 December

ECLAC (2010) The Economics of Climate Change in Latin America and the Caribbean: Stylized Facts, United Nations Economic Commission for Latin-American and Caribbean Countries, Santiago

Freeman P, Mechler R (2001) Public sector risk management in Mexico for natural disaster losses, in Innovations in managing catastrophic risks: how can they help the poor? proceedings of Wharton-World Bank conference in Washington, DC, pp 12–18

Freeman P, Martin L, Bayer J et al (2002), National system for comprehensive disaster management, phase 2: financing reconstruction, working paper, Inter-American Development Bank Regional Policy Dialogue, Washington, DC, 2 May

Grifis SH (2010) Barron's Law Dictionary, 6th edn. Barron's Educational Series, New York

IPCC (2007) Working Group III Contribution to the IPCC Fourth Assessment Report Climate Change 2007: Climate Change Impacts, Adaptation and Vulnerability, working paper, Intergovernmental Panel on Climate Change, Geneva, pp 6–11

OECD (2013) Review of the Mexican National Civil Protection System. Organization for Economic Cooperation and Development, Paris

Puente S (2005) Evaluación del fondo para la prevención de desastres naturales ejercicio 2005 (Natural disasters fund assessment), working paper, Centro de Estudios Demográficos, Urbanos y Ambientales, El Colegio de México, Mexico City

Saldaña-Zorrilla SO (2007) Socio-Economic Vulnerability to Natural Disasters in Mexico: Rural Poor, Trade and Public Response, Serie Estudios y Perspectivas CEPAL, No. 92., Economic Commision for Latin America and the Caribbean, United Nations. Mexico City

Saldaña-Zorrilla SO (2004) Mexican agriculture and its vulnerability to natural disasters and trade", in Lessons Learned from Natural Disasters Management in Developing Countries (compendium), The World Bank, Washington, DC, pp. 96-102

SHCP (2013) Informe de Rendición de Cuentas 2006-2012 (2006-2012 Accountability Report). Secretaría de Hacienda y Crédito Público (Mexican Federal Ministry of Finance) Mexican Federal Government, Mexico City

Stern N (2007) "The Impacts of Climate Change on Growth and Development". In: Stern N (ed) The Economics of Climate Change: The Stern Review. HM Treasury, London, England, p 96

UACH (2010) Evaluación externa de resultados del programa de atención a contingencias climatológicas 2010 (2010 climate hazard response program extern assessment), working paper. Universidad Autónoma de Chapingo, Chapingo

UACH (2007) Informe final de la evaluación externa del fondo para atender a la población rural afectada por contingencias climatológicas 2006-2007 (2006-2007 extern assessment final report to the fund for supporting rural population to climate hazards), working paper, Universidad Autónoma de Chapingo, Chapingo, 7 September

UACH (2006) Informe Final de la Evaluación Externa del Fondo para Atender a la Población Rural Afectada Por Contingencias Climatológicas 2005-2006 (2005-2006 Extern Assessment Final Report to the Fund for Supporting Rural Population to Climate Hazards), working paper, Universidad Autónoma de Chapingo, Chapingo, 18 September

UACH (2005) Informe Final de la Evaluación Externa del Fondo para Atender a la Población Rural Afectada Por Contingencias Climatológicas 2004-2005 (2004-2005 Extern Assessment Final Report to the Fund for Supporting Rural Population to Climate Hazards), working paper, Universidad Autónoma de Chapingo, Chapingo, 5 September

UNAM (2004) Informe Final de la Evaluación Externa del Fondo para Atender a la Población Rural Afectada Por Contingencias Climatológicas 2004 (2004 Extern Assessment Final Report to the Fund for Supporting Rural Population to Climate Hazards), Gay C, Conde C (Coord.), working paper, Centro de Ciencias de la Atmósfera, Universidad Nacional Autónoma de México, Mexico City, 4 September

World Bank (2000) Managing the Financial Impacts of Natural Disaster Losses in Mexico, Report prepared for the Government of Mexico. World Bank, Washington, DC

Chapter 5
Spatial model of incomes and migration

*En años pasados llegaron las heladas y acabaron con las
siembras en una sola noche. Y este año también. Por
eso se fueron. Creyeron seguramente que el año siguiente sería
lo mismo y parece que ya no se sintieron con ganas de seguir
soportando las calamidades del tiempo todos los años y la
calamidad de los Torricos todo el tiempo.*

JUAN RULFO: La Cuesta de las Comadres

Abstract The great human-environmental diversity of Mexico provides a
framework for an initial understanding of the wide disparities between rich and
poor. Mexico is still dominated in many regions by agriculture, and during the last
25 years, weather-related disasters have accounted for about 80% of economic
losses. This is dramatic, especially considering that this sector produces only ca. 4%
of GDP while providing a livelihood to one-quarter of the country's population.
Based on a spatial model, this chapter tests the contribution of natural disasters to
catalyzing the emigration process in vulnerable regions throughout Mexico. Besides
coping and adaptive capacity, it assessess the effect of economic losses from disasters in combination with adverse production and trade conditions during the 1990s
in triggering out-migration.

Keywords Agricultural sector • Income distribution • Trade liberalization • Assets
• Disaster prevention • Spatial dependence • Spatial econometrics

S.O. Saldaña-Zorrilla, *Natural Disasters, Foreign Trade and Agriculture in Mexico*,
SpringerBriefs in Environmental Science, DOI 10.1007/978-3-319-17359-7_5

5.1 Introduction

Migration patterns of plants and animals have been changing and adapting around the world with changes in climate conditions (Gordo *et al.* 2005,[1] Pitelka 1997[2]). In the recent history of the earth, the sensitivity of human societies to climate variations has been less than that of other species, which can mainly be explained by the widespread adoption of sedentary means of production. However, the increased occurrence and severity of natural hazards seems to be altering human migratory patterns again. Most attention paid to the impact of climate events on human migratory patterns has centered on geological-anthropological registers, e.g. Pleistocene out-migration from Africa; see McNobb, 2005,[3] and post-glaciation out-migration; see Allen *et al.*, 1999. Less attention has been placed, however, on analyzing the impact of climate and weather events on recent human migratory flows.[4] Alscher and Faist (2009), Cugusi and Piccarozzi (2009) and O'Brien et al (2008) provide evidence on the encouraging role of climate in human mobility. In this chapter, that relation is proved as well but, unlike the said works, by using spatial econometrics. Through a migration model, we test the hypothesis that recurrent natural disasters, mostly climate-related, together with decreasing income and lack of access to credit, are stimulating in Mexico. In addition, we show that migration is not a question of poverty itself. It is rather a combination of factors impacting income expectations. As the incidence of recurrent disasters and absence of credit reduce current incomes of economic agents as well as their estimates of returns of future income, they have more incentives to leave their communities. Migration is here thought of as an economic phenomenon implemented by an economic unit to respond to harmful events; in other words, a coping strategy.

This chapter presents a spatial econometric analysis which shows that recurrent natural disasters, along with lack of credit and decreasing income, stimulate

---

[1]Gordo *et al.* (2005) provide evidence that the spring arrival of long-distance migrating trans-Saharan birds is more likely to be influenced by climate conditions in wintering areas, given their direct impact on the onset of migration and its progression.

[2]Pitelka (1997) provides pre-historic and contemporary evidence on plant migration. He emphasizes the contribution of human activities through habitat fragmentation on interference with plant migration to adapt to global climatic change.

[3]Based on an archaeological study of *hominin* colonization associated with the Early-Middle Pleistocene transition, McNabb (2005) points out that climate change and especially shifts in local aridity are explanatory factors of migratory moves out of Africa. However, he concludes, the Achealean behavioral repertoire did not change much across Africa and Europe over a million years, but merely adapted to local conditions.

[4]Among the scant research on this relationship, McLemann (2006) examines the influence of natural environment on human migration and settlement patterns of rural population. He analyzes households' response in a period of adverse climate conditions in rural eastern Oklahoma during the 1930s. The focus was on those who adapted by migrating to rural California.

out-migration in Mexico and that the process spills over to nearby municipalities. We attempt to prove that migration is not a question of poverty itself, but rather of income expectations. As the effect of recurrent disasters and the absence of financing reduce economic agents' estimates of their returns on future investments, they have more incentives to leave their communities.

Given that we are looking for relevant relationships between disaster effects and emigration, we must take into account that natural disasters are geographically conditioned across regions. In other words, each municipality can be associated with both an absolute and a relative location. In terms of econometrics, the problem of absolute location is easy to solve and does not require spatial econometric techniques. This is merely a question of searching for relevant relationships between disaster effects and location characteristics. Thus, each municipality has its own set of characteristics easily estimated through the inclusion of area-specific variables such as, for instance, irrigation, distance to market or municipality, or region-specific dummy variables.

For the question of spatial dependence it is rather the relative location, or the *relative distance* between objects that is of importance for spatial econometrics. In much the same way as in time-series analysis, where what happened in previous periods influences the outcome in the current period, in spatial econometrics, a municipality's neighbors influence the outcome in a given municipality. Since influence works in both[5] directions; that is, municipality A influences municipality B and vice versa, spatial econometrics is required for estimation. Unless this is done, one might end up with both biased and inefficient estimates, as a number of authors have pointed out in, for instance, the economic growth literature. Regional growth may be dependent on growth in other regions, giving more weight to those regions which are geographically closer (i.e. Sandberg 2004; Le Gallo *et al.* 2003; Fingleton 2003; Rey and Montori 1999; Fujita and Krugman 1995; David 1984).

Hence, in our case the location of a disaster-struck municipality relative to the location of other regions/municipalities in geographical space is of interest. In addition, geographical clusters of natural disaster occurrences across regions might correlate with low-income regions in Mexico. The use of spatial econometrics is therefore vital for correct estimation and for more detail in studies of these relationships; namely identifying how clusters of regions recurrently affected by disasters and without financial measures for disaster prevention are related to clusters of low-income workers (see Anselin 1995a).

The remainder of the chapter is organized as follows. Section 5.2 discusses the nature of vulnerability to trade, production and natural disasters in Mexico. Section 5.3 presents the model, Section 5.4 discusses the results, and Section 5.5 concludes the chapter.

---

[5]Actually, the influence occurs in many directions over space among the n observations. The pattern of influence is summarised in a matrix W of spatial weights.

5.2 Vulnerability to natural and economic hazards in Mexico

Along with production limitations and trade-related stressors, the increased frequency and severity of natural hazards in Mexico has a negative impact on household income throughout the country. In light of current adverse economic conditions, recurrence of extreme climate events, and lack of affordable instruments for hedging against disaster risk, municipalities of predominantly subsistence farmers cannot recover from pre-existing marginalized conditions. They remain trapped in a vicious circle of high vulnerability, insufficient disaster management instruments, and low income. As the income expectations of economic agents turn ever more pessimistic due to these hazards which threaten their future incomes, is stimulated. During the period analyzed here, the limited capacity of the rural economy to provide employment has also significantly increased the flow of migration from the most affected regions to large cities, as well as to abroad. This leads to a complex process, avoidable only by improving current production conditions and disaster prevention.

At the national level, out-migration in Mexico has increased dramatically over the past two decades. The 1990 population census reported that 0.24% of the Mexican population was residing abroad; in the 2000 census this figure had risen to 0.41% (INEGI 2000). As currently about one-quarter of Mexico's labor force still works in agriculture, and as conditions in the countryside are continually worsening, rural workers do not have incentives to cease migrating not only to urban Mexico, but also to urban and rural USA, where today one of every three immigrants was born in Mexico (ca. 12 million). Mexicans make up the largest immigrant group in the USA as a whole, and in 31 of the 50 states of the country (Center of Migration Studies of Washington 2003; US Census Bureau 2002).

In Mexico, the agricultural sector experiences the most damage from disasters and contributes less than 4% to total GDP. However, disaster losses are critical for agricultural livelihoods and rural incomes, which involve around one-quarter of the national population. Depressed rural incomes can be attributed to the destruction of capital stock and disruption of production resulting from natural disasters along with increasing pressure from economic stressors, such as depressed agricultural prices.[6]

Although social programs in Mexico have attempted to tackle poverty, the resource flows they generate are temporary, as poverty is a structural issue caused by the working of the entire economic structure. As presented in Chapter 3, the increased frequency of natural hazards over the past 35 years in Mexico has implied increased economic losses as well. Current production structure and trade-related hazards in the countryside have contributed to stressful living conditions, amplified by insufficient

---

[6] Despite the increasing frequency and economic losses from natural disasters in Mexico, the economic impact on migration have to date not been addressed. The existing literature addressing the economic aspect of natural disasters in this country is limited to the recording of loss from disasters (Bitrán 2001), and to assessing and projecting of insurance markets (Kreimer et al 1999). Further research focuses on concrete issues, rather than providing a broader economic view.

government response to mitigate the resulting negative impacts. Nearly half the total population of Mexico lives in poverty, which is mostly concentrated in rural areas, affecting as much as 74% of the rural population (World Bank 2002).

5.2.1 Exposure to natural hazards

In Mexico, weather and climate-related events have been the most recurrent and damaging natural disasters. They are responsible for 80% of economic disaster losses during the period 1980–2005, and affect mostly the agricultural sector. As seen in Chapter 3, floods and droughts are increasingly affecting the whole country, and in some regions in the north, a process of desertification has started, with consequences for the economy. In 2005, per capita GDP agricultural growth was negative, which is mainly explained by the severe hurricane season and delays in rainfall, reducing the crop area by 34% (INEGI 2006). The population most affected are small-scale rainfed farmers of traditional crops such as maize, beans and coffee (ECLAC 2006). Over the past 15 years, the area of irrigated farmland has remained constant, covering only 24% of total cultivated area (INEGI 2013). This makes agriculture extremely vulnerable to climate and weather conditions. During the past decade, 14 million hectares of crops, of which less than 10% were insured, were lost to natural disasters (La Red 2013).

In addition, climate change is predicted to make these conditions even more challenging. Even slight variations in the climate implies very large costs in Mexico as many places are already close to the upper temperature tolerance of activities such as crop production, which demands immediate adaptation (IPCC 2007).

5.2.2 Production and trade-related hazards

Current adverse economic conditions contribute to the stress of small-scale farmers in Mexico. Between 1993 and 2004, per capita GDP growth in the agricultural sector was 0.3%, compared to 1.2% in the economy overall. As seen in Chapter 3, agricultural growth was too low to support rural population growth, while overall economic growth was too low to absorb the excess workforce from the countryside. Low-income farmers represent 67% of all farmers, but contribute only 22% to overall agricultural production. The deterioration of terms of trade in the agricultural sector, discussed in Chapter 3 as well, can explain the long-term downward trend in agricultural prices, which the current trade liberalization process dramatically exacerbates. In the short term, the comparatively lower prices of agricultural goods sold by Mexico's trade partners, most likely due to high subsidies, have resulted in higher agricultural imports, accelerating the downward trend of agricultural prices over the past 20 years. Grain imports from the USA explain the high trade balance deficit in the agricultural sector to a large degree (ca. 20% of traded value; ECLAC 2006).

Since the implementation of NAFTA, only a small proportion of farmers, namely those with access to technology, have re-oriented their production to higher-priced exportable crops. While exports of vegetables, grown mainly by high-technology farmers, have risen significantly (9.6%) during the period 1994–2004, grain exports, grown mainly by low-technology farmers, have remained practically unchanged. Grain imports, however, grew 7% during the same period (ECLAC 2006).

Based on a spatial-temporal approach, Mendelsohn *et al.* (1994) measure the economic impact of climate change on agricultural land prices in the US, and in turn on agricultural revenues. These account for the expected economic substitution made by economic agents as climate conditions change; that is, for the adoption by farmers of adaptation measures due to temperature and precipitation increases in order to shift from revenue-declining activities to those more profitable under such conditions, and so to maximize their *best-use value function*. Their estimates yield lower economic losses from climate change compared to the traditional production-function approach, as the latter leaves latent economic substitution out of the analysis. In the case of Mexico, economic substitution rarely explores in-place options, such as crop substitution, but rather tends to make use of out-migration, as more in depth analyzed in the next chapter. The production structure of agriculture in some regions of Mexico does not exploit local competitive advantages. Persistently growing grains (i.e. maize) in regions which are not climatically and environmentally suitable contributes to low productivity (Gay *et al.* 2005). In addition, these crops generally bring low prices, which means that farmers miss out on the opportunity to increase their income by changing to more profitable crops (e.g. vegetables; ECLAC 2006).

From the 1980s on, the Mexican government began withdrawing from the business of granting rural credit and marketing crops, making farmers' income uncertainty more acute. As part of an extensive series of reforms implemented over the past 20 years, agricultural policy has shifted towards less support and fewer subsidies, aiming to facilitate the emergence of a private sector-driven rural economy and largely liberalized markets. This has implied less public intervention in the agricultural marketplace through the dismantling of state-owned companies for agricultural production and marketing support. However, subsistence farmers are particularly susceptible to the negative effects of government withdrawal. Since they are often not eligible for private credit, the cost of key production and marketing services, e.g. freight, fertilizers, storage, and crop insurance become unaffordable to them (Conde et al. 2006). In addition, they have no access to technical and financial support for economic substitution. In fact, credit for the agricultural sector has decreased by 80% over the last ten years, representing only 2.5% of total credit (ECLAC 2006), and irrigation coverage—despite being increasingly required—has not been expanded since the 1980s (INEGI 2013). Finally, high transport costs and poor roads have been crucial factors in lowering the competitivity of farmers. Freight costs in Mexico are higher than in its main trade partners. For instance, the average price of one ton-kilometer of terrestrial freight is 20% more expensive in Mexico than in the USA; 0.035 and 0.029 USD/ton-km, respectively (Moreno-Quintero 2004).

5.2.3   Government intervention to reduce vulnerability

Despite the current availability of the government instruments to prevent disasters presented in Chapter 4, the strategies of public disaster management have been mainly reactive. The social benefit of the few projects implemented to reduce the risk of future disasters is still unclear. The government rebuilds not only public infrastructure, but also housing, and provides some relief to the poor after disasters. Loss sharing is based on a combination of ex-post instruments, including, for example, budget diversion and foreign credit. Risk transfer uses mainly ex-ante instruments such as insurance, mitigation, and more recently a contingency fund and a catastrophic bond. However, there is more emphasis on ex-post instruments, mainly on rebuilding, and fewer resources allocated to prevention measures. Unlike current insurance cross-subsidization in the UK (Linerooth-Bayer and Vari 2004), in the early 1990s the Mexican government started reducing the fiscal burden to farmers in low-risk areas, who in fact were subsidizing farmers growing crops in high-risk areas through flat premiums. Although the current subsidy system of crop insurance is financially much healthier than during the 1970s and 1980s, cultivated insured land in Mexico decreased from 40 to 10% between 1990 and 2000. During the same period crop damage due to natural disasters doubled. Despite the sophisticated design of current disaster mitigation instruments, few resources have been allocated to them during the past two decades. The contribution of investment to reducing vulnerability to natural disasters seems to be underestimated. In 2003, expenditure on rural natural disaster management represented less than 0.01% of spending on social programs, and less than 0.02% compared to agricultural programs in Mexico (Saldaña-Zorrilla 2006). These proportions seem to be too low if one considers the effect of natural hazards on income, which is further tested using the models of the present chapter.

5.3   The model

In the absence of financing instruments for disaster prevention and recovery, regions where natural disasters and adverse economic conditions occur more frequently tend to maintain low capital accumulation rates. This leads, all else being equal, to a reduced capital-labor ratio and in turn to a lower marginal product of labor (Barro and Sala-i-Martin 1995). Assuming that the marginal product of labor determines salaries, mean salary tends to decrease in these regions. As other regions have higher salary levels, the labor force is stimulated to emigrate, first to higher income regions within the country (Krugman and Obstefeld 2006), and then abroad if these regions are unable to fulfill future income expectations, as tested in the following chapter.

For Todaro (2000), migration is primarily an economic phenomenon, which for the individual migrant can be a quite rational decision despite the existence of urban unemployment. The Todaro model postulates that migration takes place in response to rural-urban differences in expected income rather than actual earnings. The decision to migrate is taken in order to maximize expected earnings in the individual's lifetime, and so for a given time horizon, urban areas might therefore be a rational alternative. In other words, if the damage from hazards exceeds the coping and adaptive capacity of vulnerable economic agents, they find emigration to higher-income regions to be a viable strategy, even in the face of restrictions on labor mobility and the difficulty of crossing international borders.

As migration determinants, our model includes increases in the share of population earning less than twice the minimum wage (impoverishment), access to credit, frequency of natural disasters, production structure, level of education, and distance to the closest trading center. We here emphasize the recurrence of disasters to underline the importance of repetitive disasters in influencing expectations. Empirical evidence from Mexico shows that if large economic losses affecting income are experienced in a municipality due to just one major event, households do not necessarily perceive it as a threat for the future. If the municipality is a low-frequency-disaster municipality, isolated disasters do not necessarily alter expectations of future income, as further analyzed in Chapter 6. For this reason, we use the cumulative number of reported disasters instead of monetary economic losses. Availability of credit in these regions may counteract the chain reaction triggered by recurrent disasters, regenerating expectations of future income. Access to credit is included to account for a representative (financial) coping strategy, which may counteract reductions in the capital/labor ratio following disasters and improve income expectations. We model adverse production structure through high ratios of grains (prices falling) to vegetables (prices rising), included in the variable *GRAVE*. Level of education and distance to trading centers are included in order to examine the propensity the most educated population to emigrate and the role of proximity to urban centers.

5.3.1   Sources and methods

The regression analysis was performed on data from the 2,443 municipalities of Mexico. Data sources on natural disasters are DesInventar, La Red, and CENAPRED. Data on household income, asset value, credit, geographic distances, insurance and irrigation were obtained from the National Institute of Statistics and Informatics of Mexico (INEGI), corresponding to the 1990 and 2000 agricultural and economic national census. Data on agricultural prices were obtained from SIACON, the Ministry of Agriculture and Livestock. The regression analysis was performed using the SpaceStat program, version 1.91 (Anselin 1995b).

Two kinds of spatial dependence are commonly assumed to have the potential to contaminate the analysis. The first arises when, for instance, prices of adjacent observations move together due to unobservable common or correlated variables, i.e. lack of stochastic independence between observations. This dependence leads to inefficient estimates if left unresolved. The problem is discussed at length in Cliff and Ord (1972, 1981). A partition of the error term into two parts, together with a spatial weights matrix, solves this spatial dependence problem. This model is known as the Spatial Error Model.

$$y = X\beta + \varepsilon$$
$$\varepsilon = \lambda W \varepsilon + \xi$$

The second and more serious problem of spatial dependence occurs when there is spatial correlation in the dependent variable between observations. An example of this is the growth rate in one region being influenced by growth rates in nearby regions and vice versa. Such dependence leads to estimates that are both biased and inefficient. For a further discussion of this problem, see Anselin (1988). The problem may be resolved in various ways. The most common is to include the dependent variable of the other observations on the right-hand side of the equation, lagged by a spatial weights matrix. This model is known as the Spatial Lag Model.

$$y = \rho W y + X\beta + \varepsilon$$

As usual, additional problems may occur during estimation, such as heteroskedasticity. These problems are solved by standard econometric methods. The classical estimation method for a proper model specification under the potential influence of spatial dependence is given, for instance, in Florax *et al.* (2003). The initial model is estimated by means of Ordinary Least Squares (OLS). The residuals are then used to test the hypothesis of no spatial dependence caused by an omitted spatial lag or by spatially autoregressive errors by means of two Lagrange Multiplier tests (the LM-lag test and the LM-error test), e.g., Anselin (1988) and Burridge (1980). When the hypothesis cannot be rejected (no spatial dependence is present), the results from the OLS may be used. However, in the event that the hypothesis is rejected by both tests, a new model should be constructed. The proper model is indicated by the most significant LM test. If only the LM-lag test is significant, the next step would be to estimate a Spatial Lag Model, and if only the LM-error test is significant, a Spatial Error Model.

The base model is a standard linear formulation, estimated by OLS:

$$lnM = \alpha_0 + \beta_1 \Delta LIP + \beta_2 lnCRE + \beta_3 lnNDR + \beta_4 GRAVE_i + \beta_5 lnEDUCSEC$$
$$+ + \beta_6 lnDIST_+ \varepsilon, \tag{5.1}$$

Further, equation (5.1) can be expanded in order to deal with the two kinds of spatial dependence, spatial lag dependence (equation 5.2) and spatial error dependence (equation 5.3) mentioned above.

$$lnM = \alpha_0 + \beta_1 \Delta LIP + \beta_2 lnCRE + \beta_3 lnNDR + \beta_4 GRAVE_i + \beta_5 lnEDUCSEC$$
$$+ + \beta_6 lnDIST_+ \varepsilon, \tag{5.2}$$

$$\ln M_i = \alpha_0 + \beta_{1i} \Delta LIP_i + \beta_{2i} lnCRE_i + \beta_{3i} lnNDR_i + \beta_{4i} GRAVE_i$$
$$+ \beta_{5i} lnEDUCSEC_i + \beta_{6i} lnDIST_{i+} \varepsilon_i \tag{5.3}$$

$$\varepsilon_i = \lambda W \varepsilon_i + \mu$$

$i = 1, 2, 3, \ldots, n$ (municipalities),

where W is a row-standardized first-order contiguity matrix, ρ and λ are the spatial autoregressive coefficients, and μ is a vector of i.i.d. errors with variance σ^2.

In addition to these specifications, we will also use the Spatial Durbin Model (Anselin, 2001). By re-arranging equation (5.3) it can be shown that the spatial error model is equivalent to an extended version of the spatial lag model which includes both a spatially lagged dependent variable and a set of spatially lagged independent variables (excluding the constant term). A second addition deals with spatial heterogeneity or structural change. As the assumption that a fixed relationship between the explanatory variables and the dependent variable holds over the complete dataset ceases to be tenable, heterogeneity may be present. This heterogeneity may be expressed as a different regression intercept and/or slopes for subsets of the data. Here, heterogeneity is controlled for by dividing the dataset into two subsets based on whether a municipality is marginalized or not. Fingleton (2003) and Sandberg (2004), for example, have used this approach.

Before presenting the results of the regression analysis, we give a description of the variables used in the analysis. The variable names with prefix *ln* refer to a variable which is a logarithmic transformation of the raw data.

lnM, Migration. This variable is a measure of the share of municipal population aged 12 years and over who migrated between 1990 and 2000, according to the INEGI 2000 national household survey.

ΔLIP, Low Income Population Increase. This variable represents the increase in poverty, calculated as the increase in the percentage of municipal workforce with income below twice the minimum wage between 1990 and 2000 (2000 constant prices). In fact, this variable does not reflect poverty itself, but rather deterioration in household income. Indeed, the greatest reductions in income occurred in the less marginalized regions of the country, although the repercussions on migration have been greater among the most marginalized, as shown below.

lnCRE, Credit. This variable is a measure of the number of agricultural units of the municipal total with access to credit. After natural disasters, this is crucial for rebuilding, for continuing to make use of their assets, and to cope with price fluctuation and further income uncertainty. We expect access to credit to have a negative impact on emigration, most likely due to its virtue of re-stimulating expectations of future income within the same geographical area.

lnNDR, Number of Disaster Reports. This variable is the count of natural disasters affecting a municipality over the period 1990–2000; that is, the recurrence level of disaster events by municipality. Regions with a higher frequency of natural disasters are more prone to out-migration. If the model is run with per capita economic losses from natural disasters instead, the coefficient is positive as well but not statistically significant. Indeed, these two variables measure different things. While the latter has a clear impact on income, confirmed in the low-income model, the former influences the perception of natural disasters in the community. It is verified in the following chapter, showing that in communities with comparatively low economic losses but quite frequent natural disasters, more interview subjects have plans to migrate. In contrast, in communities with higher economic losses but less frequent disaster events, there is less desire to migrate.

GRAVE, Basic Grains to Vegetables Ratio. This is the ratio of production of the five main grains to the ten main vegetables in Mexican agriculture (measured in tons). Given the structure of Mexican agriculture, grain producers are in a less competitive position, to judge by the notable decline in grain prices and the dramatic increase in imports of the same grains from the USA. High values for this variable are interpreted as weak trade competitivity in a municipality, while low values represent greater export potential, as Mexico has a clear advantage in vegetables compared to the USA and Canada (see Lederman, 2003). Given the declining trend of the relative price of grains compared to vegetables in the past two decades in Mexico (ECLAC 2006), this variable may also represent a sort of intra-branch negative terms of trade.

lnEDU, Middle School Education. This is the proportion of the municipal population 12 years and older that has completed middle school. We assume that for an individual, a higher level of education facilitates more an effective response to changing conditions, i.e. to natural disasters, price volatility, changes in the price of inputs and technological change, among others.

lnDIST, Distance to nearest trading center. This variable uses Euclidean distance as a measure of the distance between a given municipality and the geographically closest major trading center. The variable allows us to account for the effect of distance in reducing competitivity on local production due to increased transport costs and isolation from technological centers. The 50 most important benchmarked trading centers considered here are the main markets for agricultural producers, either because of their consumption or as a transfer node. Included are 25 state capitals, 12 maritime ports, the eight most important border crossings (seven to the USA, one to Guatemala), and five other large cities.

5.3.2 Model specification

We first estimate the model of equation (5.1) by OLS. The main reason for doing this is to make use of the OLS residuals to test for spatial dependence, and to provide guidance on how to proceed towards the final model specification, as mentioned above. As observed in Table 5.1, the OLS regression in column 1 yields expected parameter estimates with a goodness of fit of 25%. However, there are clear indications of spatial dependence, dominated by spatial error dependence as well as heteroskedasticity.

Table 5.1 Spatial regression results of migration, global model

Variables/tests	(1) OLS	(2) ML-Error	(3) ML-err, Het	(4) Spatial Durbin	(5) Spatial Durbin Het
ρ				0,672***	0,671***
λ		0,699***	0,698***		
Constant	0,479***	0,679***	0,666***	0,021	-0,004
ΔLIP	0,007***	0,006***	0,005***	0,005***	0,005***
lnCRED	-0,054***	-0,011	-0,010	-0,003	-0,003
GRAVE	0,000	-0,000	-0,000	0,000	0,000
lnNDR	0,099***	0,107***	0,103***	0,107***	0,104***
lnDIST	0,001	-0,007	-0,006	-0,005	-0,005
lnEDU	0,30***	0,203***	0,206***	0,191***	0,194***
W_ΔLIP				-0,003**	-0,003***
W_lnCRED				-0,030**	-0,030**
W_GRAVE				0,000	0,000
W_lnNDR				-0,084***	-0,086***
W_lnDIST				0,011	0,012
W_lnEDU				-0,006	-0,003
Gr. Variance_0			0,184***		0,181***
Gr. Variance_1			0,157***		0,155***
R2	0,25	0,15	0,15	0,44	0,43
R2adj	0,25				
Sq.corr.	0,27	0,24	0,24	0,55	0,55
LIK	-1846,84	-1372,20	-1236,15	-1349,81	-1347,80
AIC	3707,69	2758,39	2486,30	2727,62	2723,60
Sig-sq	0,27	0,16		0,16	
Heterosk. Random, (Koenker-Basset)	16,99***				
Hetersk. Marginalization, (Koenker-Basset)	4,74**				
Heterosk._Random (Spatial Breuch Pagan)		24,22***		66,82***	
Heterosk_ Marginalization (Spatial Breuch Pagan)				3,93**	
Lik. Ratio. Heterosk.			272,09***		4,02**
LM-Error	1265.83***				
LM-Error, Robust	73,81***				
LM-Lag	1247,12***				
LM-Lag, Robust	55,10***				
Common Factor (Wald)		37,46***			
Lik. Ratio. weight matrix		949,30***		883,96***	
LM on sp.Lag		18,16***			
LM on sp. Error				101,91***	

*** if prob. < 0.001; ** if prob. < 0.01; * if prob. < 0.05; blank otherwise

Since the Lagrange-Multiplier tests (LM) indicate that the spatial dependence problem may be resolved by an error correction, we begin with this correction for the model specification of equation (5.3). The regression is estimated by means of Maximum Likelihood (ML) and the results are presented in column 2. The parameter estimates mostly stay the same, as one would expect from this error correction. The log likelihood (LIK) increases significantly, but the common factor hypothesis (Burridge 1981) is rejected.

This means that the current model is misspecified and that the omission of spatially lagged variables cause spatially correlated residuals and should therefore be included (Florax et al 2003). By addressing the omitted variable bias, the WX terms help capture the influence of omitted variables whose effect would otherwise be subsumed in the error term, and they also provide detailed information about the spatial dependence present. Thus the Spatial Durbin model captures the influence of all other omitted variables that vary across space.

The spatially lagged variables consist of spatial averages of neighboring values, determined by the previously specified spatial weights matrix W of values for each observation. The parameter estimates of the spatially lagged explanatory variables are the average influence that neighbors have on the income level in each municipality. The regression results with and without correction for groupwise heteroscedasticity are included in columns 3 and 4. That is, we allow the variance to vary within two groups in order to resolve heteroskedasticity so as to decrease overall variance by allowing the variances to deviate between the two groups. The split is done on the basis of whether or not a municipality is marginalized. This categorization of the municipalities is then dichotomous. The variance for the group that contains the municipalities that are not marginalized is denoted Gr. Variance_0 in Table 5.1, and the variance of the marginalized municipalities is denoted Gr. Variance_1.

As can be seen in column 4 in Table 5.1, within-group variances are quite different between the groups. This is also confirmed by the Likelihood Ratio test (Lik. Ratio Marginalization), which rejects the null hypothesis of similar variances. Another way of seeing this is by comparing the group variances with the common variance (Sig-sq.). Compared to the results of the OLS regression model, the spatial Durbin specification clearly improves the goodness of fit, almost doubling the log likelihood value.

The spatial lag dependence parameter estimate (ρ) is positive and significant, indicating that municipalities with a high (low) degree of out-migration tend to lie clustered together and spill over to each other since they are influencing each other in income creation. Further comments on additional results are given in Section 5.4.

5.3.3   Controlling for marginalization

Now let us disaggregate the model into two types of municipalities; marginalized and non-marginalized. The marginalization index from the National Council of Population (CONAPO) serves as a category variable in this model. The composite index integrates measures including access to health, housing conditions, access to public utilities (drainage, clean water, etc), and others.[7] Its composition identifies the lack of sufficient individual and collective asset accumulation. This desegregation lets us observe structural differences between two different asset availability levels. Hence, we allow the parameter estimates for the two categories to vary across space; i.e. spatial heterogeneity. The model specification now becomes:

$$\ln M_{ij} = \alpha_{0ij} + \beta_{1ij}\Delta LIP_{ij} + \beta_{2ij}lnCRE_{ji} + \beta_{3ij}lnNDR_{ij} + \beta_{4ij}GRAVE_{ij}$$
$$+ \beta_{5ij}lnEDUCSEC_{ij} + \beta_{6ij}lnDIST_{ij} + \varepsilon_{ij},$$

(5.4)

---

[7] For details about the methodology used for the CONAPO marginalization index, see http://www. conapo.gob.mx/publicaciones/indices/pdfs/006.pdf .

Table 5.2 Spatial regression results of migration by subsets of marginalization

	Variables/tests	(1) OLS-Marg	(2) ML-Lag	(3) Spatial Durbin, Marg
NON-MARGINALIZED	ρ		0,633***	0,648***
	Constant	1,704***	1,03***	1,443***
	ΔLIP	0,001	-0,004	-0,005
	lnCRED	-0,064***	-0,014	0,010
	GRAVE	0,001	0,000	0,003*
	lnNDR	0,196***	0,114***	0,126***
	lnDIST	0,004	-0,009	-0,008
	lnEDU	-0,103*	-0,062	-0,045
	W_ΔLIP			0,002
	W_lnCRED			-0,045
	W_GRAVE			-0,003
	W_lnNDR			-0,037
	W_lnDIST			-0,016
	W_lnEDU			-0,161*
MARGINALIZED	Constant	0,263***	-0,089	-0,256**
	ΔLIP	0,006***	0,002***	0,005***
	lnCRED	-0,035***	-0,014*	0,002
	GRAVE	0,001	0,000	-0,000
	lnNDR	-0,031	-0,003	0,054**
	lnDIST	0,018**	0,010*	0,005
	lnEDU	0,326***	0,207***	0,209***
	W_ΔLIP			-0,004***
	W_lnCRED			-0,020
	W_GRAVE			0,001
	W_lnNDR			-0,115***
	W_lnDIST			0,024***
	W_lnEDU			0,018

(continued)

Table 5.2 (continued)

Variables/tests	(1) OLS-Marg	(2) ML-Lag	(3) Spatial Durbin, Marg
Chow Test:	21,90***	87,02***	112,19***
Constant	34,47***	32,73***	39,18***
ΔLIP	2,40	5,85**	9,51***
lnCRED	1,68	0,00	0,08
GRAVE	0,37	0,01	1,64
lnNDR	29,64***	12,46***	3,47*
lnDIST	1,63	4,55**	1,89
lnEDU	44,80	27,76***	21,37***
W_ΔLIP			2,69
W_lnCRED			0,43
W_GRAVE			2,76*
W_lnNDR			1,79
W_lnDIST			5,89**
W_lnEDU			3,79**
R2	0,29	0,46	0,47
R2adj	0,29		
Sc.corr		0,55	0,57
LIK	-1772,09	-1327,60	-1294,26
AIC	3572,17	2685,21	2642,52
Sig-sq	0,25	0,16	0,15
Hetersk. Marginalization, (Koenker-Basset)	0,00		
Heterosk_Marginalization (Spatial Breuch Pagan)		0,05	0,06
Lm-Error	1151,86***		
LM-Error, Robust	65,08***		
LM-Lag	1156,86***		
LM-Lag, Robust	70,08***		
Lik. Ratio. weight matrix		888,97***	801,86***
LM on sp. Lag			
LM on sp. Error		9,40***	66,80***

*** if prob. < 0.001; ** if prob. < 0.01; * if prob. < 0.05; blank otherwise

with the addition of similar spatial adjustments where necessary, as mentioned above.

$i = 1, 2, 3, \ldots, n$ (municipalities)

$j = 0$, if i is a non-marginalized municipality

$j = 1$, if i is a marginalized municipality

Table 5.2 presents the regression results. The results at the top of the table are the parameter estimates for non-marginalized municipalities, followed by the estimates for marginalized municipalities. Tests for spatial heterogeneity (the Chow test to test for overall structural stability and the stability of individual coefficients) are included at the bottom of the table. The table concludes with a section of ordinary regression diagnostics.

The test of the residuals from the OLS estimation points to a spatial lag model. The spatial lag model with adjusted groupwise heteroskedasticity is estimated by Maximum Likelihood. The regression results are given in column 3.

In terms of goodness of fit, this model explains about 47% of the variance, and the loglikelihood (LIK) value of -1293 is a clear improvement over the biased and inefficient OLS regression results, which yielded a log likelihood value of just -1772.

5.4   Interpretation of results

Regions more frequently affected by natural disasters, with relatively high increases in poverty and declining crop prices have historically had higher migration rates (Table 5.1). In addition, high levels of migration occur in regions with high education levels within marginalized regions (Table 5.2), confirming the presence of human capital drain. The classical belief that reducing excess workforce leads to increasing the marginal product of labor is a risky statement in light of these results, since the workforce remaining in the community is the least skilled. In addition, distance proves not to be a real obstacle to migration, given that most migration originates in more isolated communities (Table 5.2).

The positive sign of ΔLIP in Table 5.1 suggests that municipalities with a workforce that has experienced larger reductions in wages are more prone to out-migration. This suggests that the greater the **poverty increase** during the 1990s, the greater the migration rate between 1990 and 2000. From Table 5.2 we may conclude that this effect operates only in marginalized municipalities. Hence, out-migration occurs significantly more often in marginalized municipalities as a result of an increased proportion of low-income population than in non-marginalized municipalities. Out-migration is further influenced by factors in the surrounding municipalities; increased poverty in neighboring municipalities of municipality i inhibits migration from i, (W_ΔLIP).

Credit is not statistically significant in its effect on migration in the overall model (Table 5.1). As with insurance, the availability of agricultural credit decreased dramatically (ca. 80%) during the 1990s, making hard to measure its impact on income. The sign on the coefficient of credit in marginalized regions is positive, although insignificant (Table 5.2, column 3). We have reason to assume that lack of credit would stimulate the workforce to migrate, given that they recur directly to out-migration when faced with adverse conditions, as they cannot count on credit as a feasible coping strategy.

We observe, however, a significant negative coefficient for the spatially lagged variable W_lnCRED in Table 5.1. This means that having neighbors with comparatively higher access to credit helps to keep migration rates lower. As credit conventionally facilitates a means for increasing the risk pool, helping to smooth the negative effects of hazards, a positive spillover effect from neighbors with low losses may facilitate obtaining jobs nearby, and lessen the chances of immediately recurring to out-migration.

Our results show that a high **disaster frequency** encourages more people to move away from their municipalities. All else being equal, a 10% increase in disaster frequency raises migration rates by 13% in non-marginalized regions, and 5% in marginalized ones. Factors in the surrounding municipalities limit migration, as shown by the prevailing negative coefficients of the corresponding spatially lagged exogenous variables in Table 5.1. For instance, high disaster frequency in surrounding regions, W_lnNDR, reduces the probability that people will migrate. It is most likely due to the catch-up increases in crop prices that may be set by less-affected producers due to increased scarcity.

A higher frequency of natural disasters tends to make the non-marginalized population comparatively more prone to migrate, perhaps due to their higher educational levels, making them better informed about emigration as a coping strategy in the absence of credit and other alternative financing sources.

Although the ratio of **grains to vegetables** (GRAVE) provides negligible results for the overall model, the marginalization partition suggests that those non-marginalized regions growing predominantly grains tend to have increased migration levels (Table 5.2). This is most likely due to their greater involvement in trade, while the most marginalized municipalities rely more on self-consumption. All else being equal, a decrease of 10% in the grain-vegetables ratio in the less marginalized municipalities leads to a 3% decrease in out-migration.

As observed, **education** is positively correlated with out-migration (Table 5.1). However, as shown in Table 5.2, the results for non-marginalized municipalities are no longer significant. For the marginalized municipalities, education is still significant for explaining out-migration. This means that among marginalized municipalities, migrants come from regions with a comparatively higher mean level of education. This also suggests that less educated people are less prone to implement migration as a way to cope with hazards either because they are more averse to changing their residence, have less means to do so, or have a more resigned attitude, given hierarchical governmental organization prevailing in the recent past, which has a greater effect on the marginalized population, and is still present in many regions. The prevailing fatalistic way of life of most subsistence farmers with low levels of education, abundantly expounded in Thompson *et al.* (1990), provides further evidence for these results.

High out-migration rates in marginalized municipalities occur if they have experienced a notable increase in their low-income population combined with an initial high education level. In the spatial Durbin model, having a greater proportion of educated people in non-marginalized municipalities reduces migration. This could be because their more educated neighbors generate more economic activity and hence jobs in the region, or simply that the opportunities at home are already sufficient.

It was expected that marginalized people near big cities would be more likely to migrate as their **distance** and transport costs to large cities are lower. This is not the case, however, since none of the models present any significant parameters for this factor, except for the spatially weighted distance variable for marginalized municipalities (Table 5.2), which has a positive significant parameter. One interpretation may be that marginalized households are unable to find jobs in nearby municipalities that are far from the market, and a solution to find employment would then be to migrate a greater distance.

In summary, lower income, disaster frequency, and education proved to determine migration with much better results than any other variables in the overall model. In general terms, the prototype of an emigration-prone municipality in Mexico has the following combination of characteristics: substantially dropped income over the past ten years, frequently affected by natural disasters, marginalized, relatively more educated, grain-intensive farming, located far from larger urban centers, and low access to credit.

Conclusions

As weather and climate-related disasters are becoming more frequent and destructive in Mexico, the model suggests that out-migration will continue to rise in the coming years if an active disaster prevention strategy and structural adaptation measures are not duly implemented. Municipalities with higher emigration rates were those with greater income decreases and increasingly affected by recurrent natural disasters during the 1990s. The model suggests that out-migration may be a result of future income expectations. Migration is therefore a rational decision made by economic agents based on estimated future returns on their assets, accounting for expected net assets and income losses after internalizing available financial resources. The model also suggests that increasing access to agricultural credit may reduce the stimulus to migrate as it reduces the negative impacts of hazards in the affected region. The population segment most significantly prone to migrate is that of those who are marginalized, have higher levels of education, and live far from trading centers. As the current transportation infrastructure does not support them sufficiently to enable them to compete as economic agents, they are utilizing it instead to migrate.

The temporal-geographic identification of out-migrating regions in this study may serve as a baseline for government decision-making on investment projects for expanding crop insurance coverage, access to credit, and promoting more public infrastructure investment in areas identified as vulnerable by climate change scenarios. Finally, current investment in education will continue to yield benefits merely as an instrument for training potential migrants as long as reforms to enable the accumulation of assets and consequently to reduce social polarization do not materialize in this country.

Chapter References

Allen et al. (1999) Rapid environmental changes in southern Europe during the last glacial period. In: Nature 400(6746):740–743

Alscher S, Faist T (2009) Environmental factors in Mexican migration: the cases of Chiapas and Tlaxcala. Case study report on Mexico for the Environmental Change and Forced Migration Scenarios Project. http://www.each-for.eu/.

Anselin L (1995a) Local Indicators of Spatial Association. Geographical Analysis 27:93–115

Anselin L (1995b) SpaceStat, A software program for the Analysis of Spatial Data, Version

Anselin L (1988) Spatial Econometrics: Methods and Models. Kluwer Academic. Dordrecht

Anselin L (2001) Spatial econometrics. In Baltagi, Badi, editor, A Companion to Theoretical Econometrics, pages 310–330. Blackwell, Oxford

Barro R, Sala-i-Martin X (1995) Economic Growth. McGraw Hill, New York

Bitrán et al. (2001) Características del impacto socioeconómico de los principales desastres ocurridos en México en el periodo 1980–1999. Centro Nacional de Prevención de Desastres, Secretaría de Gobernación. Mexico City

Burridge P (1981) Testing for a common factor in a spatial autoregressive model. Environment and Planning A 13:795–800

Burridge P (1980) On the Cliff-Ord test for spatial correlation. Journal of the Royal Statistical Society B 42, 107–108

Center of Migration Studies of Washington (2003) Population Growth, Immigration, and the Problem of Sprawl. August 2003. By Roy Beck, Leon Kolankiewicz, and Steven A. Camarota

Cliff A, Ord K (1981) Spatial processes: models and applications.Pion, London

Cliff A, Ord K (1972) Testing for Spatial Autocorrelation Among Regression Residuals. Geographic Analysis, 4

Conde C, Ferrer R, Orozco S (2006) Climate change and climate variability impacts on rainfed agricultural activities and possible adaptation measures. A Mexican case study. Atmósfera

Cugusi B, Piccarozzi M (2009) Environmental change and human mobility. A thematic literature and organizational review. Paper prepared for the International Conference "Climate Change and Human Mobility in Africa: Dialogue for a strategic cooperation between Italy and Africa", Rome, 21st of April 2009

David PA (1984) High Technology Centers and the Economics of Locational Tournaments. Mimeo, Stanford

ECLAC (2006) México: Crecimiento Agropecuario, TLCAN, Capital Humano y Gestión del Riesgo. United Nations Economic Commission for Latin-American and Caribbean Countries, Mexico City

Fingleton B (2003) Models and Simulations of GDP per Inhabitant Across Europe's Regions: A Preliminary View. In: European Regional Growth. Ed: B. Fingleton. Pp 11–53. Springer-Verlag

Florax R, Folmer H, Rey SJ (2003) Specification Searches in Spatial Econometrics: The Relevance of Hendry's Econometric Methodlogy. Regional Science and Urban Economics 33:557–579

Fujita M, Krugman P (1995) When is the Economy Monocentric? Von Thunen and Chamberlin Unified. Regional Science Urban Economics 25(4):505–528

Gay C, Estrada F, Conde C et al (2005) Impactos Potenciales del Cambio Climatico en la Agricultura: Escenarios de Producción de Café para el 2050 en Veracruz (Mexico). In: Rasilla D, El Clima, entre el Mar y la Montaña. Universidad de Cantabria: 651–660. Santander

Gordo D et al (2005). Do changes in climate patterns in wintering areas affect the timing of the spring arrival of trans-Saharan migrant birds? In: Global Change Biology. Volume 11, Issue 1, January 2005, Pp. 12-21

INEGI (2013) National Institute of Statistics, Geography and Informatics. http://www.inegi.gob.mx Accessed 3 july 2013

INEGI (2000) Anuario de Estadísticas por entidad federativa. National Institute of Statistics, Geography and Informatics. http://www.inegi.gob.mx/difusion/espanol/bvinegi/aef/indef.pdf Accessed 3 may 2002

INEGI (2006) National Institute of Statistics, Geography, and Informatics. On-line consultation on May 6, 2006: http://www.inegi.gob.mx

IPCC (2007) Climate Change Impacts, Adaptation and Vulnerability, Working Group III, Contribution to the IPCC Fourth Assessment Report Climate Change 2007. Intergovernmental Panel on Climate Change

Kreimer A et al. (1999) Managing Disaster Risk in Mexico: Market Incentives for Mitigation Investment. The World Bank Disaster Management Facility. Washington, DC

La Red (2013) Social Studies Network for Disaster Prevention in Latin-America. Panama. www.desinventar.org Accessed 21 December 2013

Le Gallo J, Ertur C, Baumont C (2003) A Spatial Econometric Analysis of Convergence Across European Regions 1980–1995. In: European Regional Growth. Ed: B. Fingleton. Springer-Verlag, pp 99–129

Lederman D et al (2003) Lessons learned from NAFTA. NAFTA's remaining trade barriers. The World Bank Group, Washington, DC

Linerooth-Bayer J, Vari A (2004) A Model-based Stakeholder Approach for Designing Disaster Insurance. In: The Role of Local Governments in Reducing the Risk of Disasters, K. Demeter, N.E. Erkan, A. Guner (eds), The World Bank Institute, Washington, DC, USA, pp. 159–185 [2004]

McLemann (2006) Migration out of 1930s rural eastern Oklahoma: Insights for climate change research. Great Plains Quarterly 26(1):27–40

McNabb J (2005) Hominins and the Early-Middle Pleistocene transition: Evolution, culture and climate in Africa and Europe. In: Geological Society Special Publication. Issue 247, pp. 287–304. Southampton, UK

Mendelsohn R, Nordhaus W, Shaw D (1994) The Impact of Global Warming on Agriculture: A Ricardian Analysis. American Economic Review 84:753–771

Moreno-Quintero E (2004) El sobrepeso en el transporte carretero de carga. In: Planner-User Interactions in Road Freight Transport. PhD Dissertation. Leeds University

O'Brien K, Sygna L, Leichenko R et al (2008) Disaster risk reduction, climate change adaptation and human security. A commissioned report for the Norwegian Ministry of Foreign Affairs. GECHS Report 3

Pitelka L (1997) Plant migration and climate change. American Scientist. Sept.-Oct., 1997. Vol. 85, No. 5. pp. 464 (10). The Scientific Research Community. Gale Group

Rey SJ, Montori B (1999) US Regional Income Convergence: A Spatial Econometric Perspective. Regional Studies 33(2):143–156

Saldaña-Zorrilla, SO (2006). Reducing Economic Vulnerability in Mexico: Natural Disasters, Foreign Trade and Agriculture. Dissertation, Wirtschaftsuniversitaet Wien. Vienna, Austria

Sandberg K (2004) Hedonic Prices, Economic Growth, and Spatial Dependence. Umea Economic Studies No. 631. Pp. I, 1-12; II, 8-15. Umea University. Umea, Sweden

Thompson M, Ellis R, Wildavsky A (1990) The Civic Culture Reexamined. In: Cultural Theory. Political Culture Series. Westview Press. Pp. 252–254. Oxford, UK

Todaro MP (2000) Urbanization and Rural-Urban Migration: Theory and Practice. In: Economic Development. Seventh Edition. Addison Wesley Longman. Essex

US Census Bureau (2002) http://www.census.gov/acs/www/ Accessed 12 February 2002

World Bank (2002) Country assistance strategy of the World Bank Group for the Mexican United States, Washington, DC

Chapter 6
Stakeholders' views in reducing vulnerability and resilience building

De los ranchos bajaba la gente a los pueblos; la gente de los pueblos se iba a las ciudades. En las ciudades la gente se perdía, se disolvía entre la gente. «¿No sabe ónde me darán trabajo?» «Sí, vete a Ciudad Juárez. Yo te paso por doscientos pesos... »

JUAN FULFO, *Paso del Norte*

Abstract This chapter examines how climatic events affect agricultural livelihoods. Special emphasis is given to the effects of natural disasters on migration patterns. In addition, it aims to assess policy options to reduce the vulnerability of small-scale farmers (e.g. government-supported insurance schemes). To further this aim, this work draws on stakeholder consultations and descriptive analysis in three communities in the southern state of Chiapas, Mexico. It also puts forward stakeholder-based solutions, which embrace loss-sharing and risk-transfer mechanisms. The coping strategies revealed in this study encompass both immediate responses (e.g. sources of off-farm income, post-disaster financing sources, and emigration plans), and more structural and long-term strategies, such as re-orientation of production and improvement of infrastructure for production.

Keywords Vulnerability • Resilience building • Rural poverty • Subsistence farmers • Development • Natural disasters • Trade • Adaptation • Productive reorientation • Household survey • Rural livelihoods • Coping strategies • Community adaptation

6.1 Introduction

In Mexico, rural poverty is a complex issue with deep-rooted causes such as the education system. It is being increasingly exacerbated by recent changes in economic, policy and dynamic processes (such as weather-related disasters). Most Mexican farmers live in poverty and are vulnerable to an increasing number of

© The Author 2015
S.O. Saldaña-Zorrilla, *Natural Disasters, Foreign Trade and Agriculture in Mexico*,
SpringerBriefs in Environmental Science, DOI 10.1007/978-3-319-17359-7_6

natural and socio-economic hazards. This helps to explain why agriculture produces only 4% of Mexico's GDP despite employing around one quarter of the national labour force. Analysing the relationships between rural poverty, natural disasters, trade and agricultural policies at national, regional and community levels is crucial for a comprehensive assessment of vulnerability. The increasing economic losses from natural disasters in this country have led to the design of a very novel public response, whose prevention measures, however, fail in reaching the most vulnerable of society. The budget decentralization in Mexico from the 2000's has made municipalities responsible for identifying disaster risk as a condition of obtaining financing from the federal government for prevention and mitigation measures. In many cases, however, municipalities have tended to underestimate the relevance of natural disasters, preventing them from conducting risk analyses and, in turn, impeding them from obtaining federal financing, as discussed above in Chapter 4. The present research found community leaders to be not fully committed with these issues even in recurrently affected areas. Government intervention consists mainly of post-disaster aid which, while helpful, is insufficient and usually late. As a result, the vulnerable population, especially in rural areas, must cope with the detrimental effects of natural hazards by themselves. These effects often far exceed the communities' coping and adaptive capacity, contributing to increased poverty and emigration.

This chapter presents empirical results derived from a ten-year survey (2004-2014) in Southern Mexico on vulnerability to hazards. It highlights coping strategies and the complex processes underlying rural vulnerability, and identifies priorities for action. The emphasis in this survey is on capturing a spectrum of coping strategies as responses to the negative effects of natural disasters in a framework of trade liberalisation and limited government support for agricultural production, commercialisation, and financial services in Mexico. Farmers' coping and adaptation strategies are highly conditioned by past trade policy decisions, in some cases even amplifying the negative effects of natural disasters and obstructing farmers as they attempt to overcome poverty. The chapter is organized as follows. Section 6.2 presents a brief conceptual discussion of vulnerability as well as includes a background for understanding the multidimensional nature of rural vulnerability in Mexico. It discusses the roles of trade, natural disasters, rural poverty and historic land tenure in contributing to the vulnerability of the farmers in the region under study. Section 6.3 summarizes the survey methodology, and presents the profiles and hazard exposure of the selected communities. Section 6.4 presents the main findings on the farmers' coping and adaptive strategies derived from the field work in 2004-2005, further complemented by a discussion of suggestions in Section 6.5. Section 6.6 discusses the leaders' and authorities' opinions about how vulnerability may be reduced. Section 6.7 provides the main results from a second field work conducted in 2014, which compares the vulnerability and resilience conditions of these communities ten years after. Section 6.8 presents the conclusions.

6.2 Multi-dimensional nature of farmers vulnerability in Mexico

Given increasingly adverse climatic conditions in Mexico, which are likely to be related to climate change (Conde and Gay 1999), along with uncertain rural incomes resulting from insufficient agricultural infrastructure, market uncertainty and price volatility (Eakin 2003), subsistence farmers are highly vulnerable and have increasingly been adopting off-farm coping strategies, including emigration and informal urban employment (ECLAC 2001). These are not necessarily viable long-term solutions to the problem, but are immediate relief measures. There is thus a need for investigating feasible vulnerability-reduction strategies that take these hazards into account in a comprehensive manner as well as for resilience building (Norris et al 2008).

If current agricultural free trade conditions are maintained in Mexico, prices will likely remain depressed. In 1986, Mexico began a process of trade liberalization. First with the adherence to the General Agreement on Tariffs and Trade (GATT) in 1986 (current World Trade Organization, WTO), and increasingly following the implementation of the North America Free Trade Agreement (NAFTA), agricultural prices have decreased dramatically, especially those of maize, the basis of Mexican agriculture. While low prices are an advantage to consumers of maize, they have decreased most farmers' incomes given their technical and financial inability to change to more profitable crops (Yúnez-Naude and Barceinas 2003).

As seen throughout the previous chapters, natural disasters have been increasingly causing damage to Mexican agriculture, especially to subsistence farmers. Significant natural hazards include droughts in the north of the country and flooding in the south. Weather-related events have caused 44% of total losses over the period 2000–2005, and crop agriculture has been the sector hardest hit by these events. Data from the National Institute of Statistics (INEGI 2005) indicate that in Mexico in 2002 over two thirds of total cropland was rain-fed. This shows the high dependence of agricultural production on meteorological phenomena. At the same time, the government no longer grants or manages rural credit, crop marketing services or crop insurance as it did up to 1988. Private companies have now replaced public institutions in most of these roles, but the majority of farmers cannot afford these services, nor can they afford credit or crucial inputs such as seed and tractors. In addition, market asymmetries prevent farmers from integrating into the production chain, denying them access to, for example, storage and shipping. After the government withdrew from providing direct crop insurance services to farmers (Hernández-Trujillo 1997, Wodon and Velez 2000), the total area of insured crops reduced dramatically. As seen in Chapter 4, during the 1980s, 40% of cropland was insured, while during the 2000s only 10% was insured.

As pointed out by Linerooth-Bayer and Vari (2004), the lack of a risk-sharing mechanism (i.e. cross-subsidization insurance from better-off farmers cropping in low-risk areas), along with insufficient incentives to promote loss-reduction incentives (i.e. risk mitigation projects) contributes toward amplification of the negative consequences of extreme events. Subsistence farmers are directly affected by disasters, which cause them net loss in capital stock and income flow. This is costly to the government, which then has to spend repeatedly on relief to the poor and on rebuilding infrastructure. This makes less funds available for public investment in the agricultural sector as a whole.

Poverty is another crucial element of vulnerability in rural Mexico. The *World Bank 2004 Report on Poverty in Mexico* estimated the poverty rate in Mexico at 52% (World Bank 2004). Rates of both moderate and extreme poverty in Mexico are high in rural areas. 74% of the rural population lives in poverty compared to 36% in urban areas (World Bank 2002). 20% of the Mexican population is considered to live in extreme poverty,[1] of which 65% live in rural areas (World Bank 2004). In addition, the livelihood of 72% of the extreme poor in Mexico is based on agriculture; Low levels of education are also much more prevalent in rural areas - 73% of the rural extremely poor have no education or less than six years of education, compared to 51.3% of the urban extremely poor (World Bank 2004).

Inequity is high in rural areas and has increased during the last decade. Using the Gini coefficient,[2] the coefficient of rural expenditure increased from 0.41 to 0.48 between 1992 and 2002 (World Bank 2004). Also, despite the fact that over 25% of the Mexican labour force works in agriculture, the sector includes only 2% of the workers who have public health insurance and social security coverage (INEGI 2005). As in other Latin American countries, poverty in Mexico is a complex issue with deep-rooted historical causes (e.g. land tenure) but is also explained by dynamic factors and processes (e.g. economic policy and external shocks). The former corresponds to what is termed 'original endowment'[3] in the form of initial allocation of wealth—after the Spanish conquest in the case of Mexico. Factors such the industrialization model employed during the 1950s and 1960s and the economic liberalization of the 1990s provide additional elements for understanding the increase of poverty in the countryside.

As discussed in Chapter 2, the typical landholding in the Mexican agrarian system is the *minifundio*, which consists of properties too small to provide those who work them with a livelihood not much above the bare minimum for survival. Along with other hazards discussed here, this *atomisation* of property prevents farmers from increasing their scale of production to realize savings, or reinvesting to upgrade technology. For that reason, holders of *minifundios*—the majority of Mexican farmers—tend to provide seasonal labour to *latifundios* (large farms) or even to other labour markets. In summary, only 45% of the total income of rural workers in Mexico is provided by agriculture, and the proportion has been decreasing for the last 30 years (ECLAC 2001).

Other factors complementing our understanding of current poverty and social inequality in the Mexican countryside include: (*i*) policies giving preference to the

---

[1] Extreme poverty is defined as living below the food-based poverty line.

[2] The Gini coefficient is a measure of income inequality. It consists of a number between 0 and 1, where 0 means perfect equality (everyone has the same income) and 1 means perfect inequality (one person has all the income, everyone else has none).

[3] For further details about Endowment, and Endowment Effect, see John List (University of Maryland) and Daniel Kahneman (*Homo sapiens vs. Homo economicus*).

manufacturing sector over the agricultural sector during the period 1940–1980 (Figueroa 1991); (*ii*) the accelerated withdrawal of the government from subsidizing, granting credit, and from investing in agricultural infrastructure and technology development since the 1980s as part of a broader economic liberalization process (see Montserrat and Chavez 2003 and Hernández-Trujillo 1997), and; (*iii*) the 60% net drop in mean weighted agricultural prices (constant 2002) over the past 25 years, due mostly to trade liberalization, as seen in Chapter 3.

Box III: INMECAFE, a dismantled coffee support institution

As in other important coffee producing countries like Brazil, Guatemala, El Salvador, Colombia, Uganda, Kenya and Indonesia, coffee regions in Mexico are located in highly marginalized regions. In Mexico, coffee is produced in twelve of the 31 states. Coffee plantations cover over 700,000 hectares, making Mexico the sixth largest coffee producing country in the world. The main producing states are Chiapas, Veracrúz and Oaxaca (SIACON 2005), where more than 3 million people rely on coffee production as main livelihood (Emanuel and Greenberg 2000).

When coffee international prices started to soar during the 40s, the Mexican federal government implemented a strategy endeavored to encourage coffee exports. As part of these efforts, the government established the Mexican Institute of Coffee, INMECAFE (Instituto Mexicano del Café), a state owned company aimed to act as intermediary between producers, exporters, and merchants. During the 70s, the government tried to reduce the excessive debt burden of small-scale farmers to coffee merchants, and INMECAFE began by purchasing coffee directly from farmers, though it was unable to grant farmers with sufficient credit in time as coffee merchants did.

The liberal economic policy of the administration of Carlos Salinas (1988–1994) led to the restructuring and further dismantling of INMECAFE in 1989, what left two options to small-scale coffee growers: either marketing their coffee though the local intermediaries (coyotes) or participating in small cooperatives, which failed in supplying them with credit and commercialization services. Currently, as most of Mexican coffee growers have no means to transport their production to export centers, they are forced again to make use of networks of local and regional intermediaries (Emanuel and Greenberg 2000).

In fact, the government in 1989 argued that the dismantling of INMECAFE was a decision based on cost-benefit analysis, which found negative net benefit from cropping coffee and onerous for the government, suggesting re-orientating local production. However, the latter has not been yet implemented, bringing some of the negative effects pointed out by farmers' leaders and local authorities in this survey.

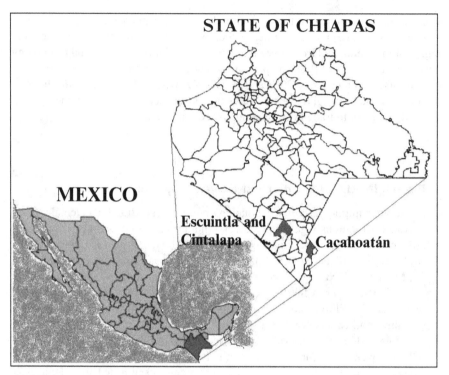

SOURCE: Saldaña-Zorrilla 2008.

Fig. 6.1 Map of Mexico with surveyed municipalities

6.3 Study sites, research methods, hazards and local conditions

The communities included in this study were Cacahoatán, Escuintla and Cintalapa, located on the coast of Chiapas, the poorest state in Mexico. The region is located near the Pacific Ocean coast bordering Guatemala (see Fig. 6.1). These communities were selected because of their history of natural disaster events combined with their conditions of extreme poverty.

While the mean number of natural disaster per municipality during the period 1990–2000 was 1.1 at the national level and 1.7 for the state of Chiapas, the selected communities are located in municipalities with means of 3.3 (Cacahoatán) and 8.3 (Escuintla and Cintalapa). According to data from La Red (2002), CENAPRED (2001) and this study's interviewees, the region has been highly affected by excess precipitation, and the most recurrent natural disasters have been (1) heavy rains; (2) winds; (3) hurricanes, and (4) landslides. The main crops in the region are coffee, maize and banana, which have all experienced dramatic price reductions during the past 15 years (ECLAC 2006, Lederman et al 2003). According to the Mexican

National Council of Population, the three communities surveyed are classified as highly marginalized (CONAPO 2004). 42% of the population in Cacahoatán and 50% in Escuintla and Cintalapa live in extreme poverty according to estimates from the Ministry of Social Development (SEDESOL 2005).

Inhabitants of these three communities rely predominantly on subsistence agriculture. According to the questionnaire data, the mean size of these farm properties is 5.2 hectares (95% confidence interval 1.59 to 8.75), the mode is 0.1 hectares, and 75% of farmers interviewed own less than 5 hectares. The farms are characterized by low technology usage, scarce access to credit and insurance, and reliance to a large extent on mono-crop agriculture. Both trade and climatic conditions affect agricultural income, but climatic conditions are most important in subsistence farming. Compared with commercial farmers, subsistence farmers are less vulnerable to unfavourable trade conditions as they rely on trade to a lesser extent. At the same time, they are relatively more vulnerable to natural disasters. Subsistence farmers rely mostly on consumption of their own crops. Losing their crops to events such as natural disasters therefore directly threatens their food supply.

To investigate the vulnerability to natural hazards of farmers in Southern Mexico, this study draws on both qualitative and quantitative research methods. Qualitative methods consist of a literature review addressing the main stressors to the Mexican economy as a whole and, further, downscaling to the surveyed region, in order to prepare the field work. Quantitative methods include, prior to travelling, community profiles preparation based on meteorological indicators, natural disasters registers and data analysis of poverty, local production and trade, among others, as well as, after travelling, data processing and statistic analysis of questionnaires. Interviewees are classified into two main categories: farmers and authorities. Interviewed farmers were agricultural producers of subsistence - those that provide enough food for the farmer and his family but not enough surpluses for regularly trading or to make significant investments to enlarge production scale. Interviews with authorities include community leaders, farmer union chairs, and governmental authorities. The questionnaire survey was conducted between October 2004 and January 2005 in three rural communities in the state of Chiapas.

The farmers' questionnaire used in this survey was conducted in Spanish language and asked both multiple-choice and open-ended questions. Responses were anonymous. The six-page questionnaire consisted of 35 questions (see questions in Appendix 1). Two-thirds of the questionnaires were distributed to groups of farmers as part of 2 focal forums in the city hall auditoriums of Cacahoatán and Escuintla. The rest were distributed house by house, where the questionnaire was just left and two hours later picked up. The average time to answer the questionnaire was 40 minutes. A total of 151 questionnaire surveys were completed in these three communities, namely 66 questionnaires were applied in Cacahoatán, 47 in Cintalapa, and 38 in Escuintla. It represents at least 15% of the population of subsistence farmers in the respective communities; mostly small land holders (mean 5 ha per household). Over 80% of surveyed farmers have possession of the land on a communal basis (*Ejido*). These households rely heavily on self-consumption agriculture. The mean age of the surveyed population is 43.7 years; 80% are 57 years old or younger. 72% are male and 26% female. Age, gender and land size distribution of the respondents are representative of the municipal distribution of workforce[4] with a 95% confidence interval.

In addition to the 151 farmers, a total of 25 authorities were interviewed. Community leaders, farmer union representatives and local authorities of this region were interviewed face-to-face and were asked semi-structured questions with similar content to those asked to farmers. Interviewed authorities embrace mainly community and municipal authorities, but also includes some leaders, authorities and policymakers at the state and federal level in order to be consistent with the overall multi-scale approach of this research (see list of interviewees in Appendix 2). These interviews were tape recorded and transcribed and the findings from these interviews are discussed in Section 6.6.

The main purpose of the survey was to assess the stakeholders' perception of environmental and economic change, to analyse their preferred coping strategies, to investigate the scope and role of government intervention in dealing with these hazards, and to find potential stakeholder-based solutions to reduce their vulnerability. From the whole survey, this chapter focuses on the following core questions: Do farmers consider that natural disasters are significantly affecting their incomes? What are the current roles of the government, private markets (e.g. insurers), farmers and other stakeholders in coping with adverse outcomes from natural disasters and income uncertainty of subsistence farmers? Other factors aside, are disasters influencing farmers' desire to emigrate? In the perception of the farmers, how can the government contribute to reducing agricultural and disaster vulnerability, including public transfer schemes (poverty reduction and support for production) and risk transfer and loss-sharing schemes?

The open questions included in the survey investigate the need for and viability of implementing crop insurance, mitigation measures and re-orientation of production to reduce vulnerability to disasters given uncertain rural incomes. These open questions were also extremely valuable in providing insights into failures and suggestions for public policy that might improve conditions for earning a livelihood. The results of this questionnaire also show how some characteristics of both policy and politics have played against vulnerability reduction in this region. Data from the 151 questionnaires was processed and analysed using SPSS software. The following sections discuss their main results.

Historical records show recurrent disasters in the region, especially during the last third of the year. In 1998 and 2005 Escuintla and Cintalapa suffered heavy rains and flooding, leaving thousands of victims and millions of dollars in economic losses (MunichRe 2005, CENAPRED 2001). Cacahoatán has suffered fewer human casualties but higher production losses from frequent heavy rains which occur once a year on average. According to reports from La Red (2002), all three communities have been experiencing unusual climatic conditions - drought in coastal areas and heavy

---

[4] Municipal distribution of age, gender and land size of workforce is based on INEGI (2005) and CONAPO (2004) registers.

rains in the mountains (Sierra Madre de Chiapas) - which are becoming more extreme. The communities are characterized by a warm humid climate with rains during most of the year in coastal areas and are usually affected by hurricanes in summer and autumn and by the summer-autumn rainy season in the mountains. In October 2005, as a consequence of Hurricane Stan, nearly 600,000 people on the Chiapas coast were directly affected by flooding and sudden river overflows. Economic losses were particularly high in Escuintla but also severe in Tapachula, Motozintla and Huixtla (see CENAPRED-ECLAC 2006, Guy Carpenter 2006 and Munich 2005).

6.3.1 Local conditions in the study sites

While the factors - foreign trade, rural poverty and historic land tenure - discussed in Section 6.2, affect the vulnerability of farmers in the entire Mexican region and therefore affect the farmers in the study sites as well, some local conditions of the study sites in addition to these factors also affect the vulnerability of the farmers. These local conditions are discussed in this section.

The prevailing mono-cropping practice increases local farmers' exposure to hazards. The *standard model of trade,* based on the postulates of David Ricardo of 1817, emphasizes the positive effect of mono-cropping which includes the exploitation of local comparative advantages, economies of scale, and benefits resulting from high specialization in Adam Smith's sense of the *division of labour* (Krugman and Obstfeld 1997). Nevertheless, mono-cropping incurs risk when the respective commodity markets turn highly volatile, prices drop dramatically, or simply because of the negative long-term *terms of trade* of agricultural goods relative to manufacturing. As Benson and Clay (2002) point out, mono-crop economies face vulnerability not only from natural disasters, but also from uncertainty in world trade due to non-diversified export products of primary commodities. The main agricultural products grown by farmers in the sample are maize (46%) and coffee (41%). The swift drop in maize and coffee prices observed during the last 25 years in Mexico tell of the net loss experienced by these farmers as agricultural yield and financing increased more slowly for a given land area.

6.4 Analysis of farmers' coping and adaptive strategies

If Mexico has programs and funds available to reduce disaster vulnerability, why are they not implemented in the study region despite high exposure? In the following sections we show how the answer lies in the lack of state government coordination to apply for funding for preventive measures, together with insufficient organization among local leaders to demand them. In addition, neither local authorities nor leaders in this region have a contingency fund or plans to establish one.

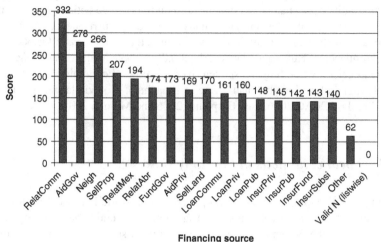

SOURCE: Saldaña-Zorrilla2008

Fig. 6.2 Importance of disaster-financing source

6.4.1   Sources of post-disaster financing

In these communities, ex-post financial instruments to cope with natural disasters predominate. These basically consist of relatives living in the same community and are complemented by post-disaster aid from the government. The questionnaire asked farmers about the importance of different financing sources for natural disasters, including public sources for risk reduction. As shown in Fig. 6.2 below, "Relatives in the community" is the most important post-disaster financing source, with the highest score (332), followed by government aid (270) and neighbour solidarity (266). Less important is the sale of property, government funds, sale of land, and community loans. Finally, the four types of available insurance —private, public, *Fondos* and subsidized— are ranked last.[5]

6.4.1.1   Production re-orientation

The *theory of Terms of Trade* expounded in Chapter 2 provides valuable elements for understanding the importance of adverse foreign trade conditions on vulnerability in the agricultural sector: if prices of the country's exports are falling relative to the price of the products it imports, it must sell that much more of its export product

---

[5] Consider that no major natural disaster had affected this region between 2000 and the application of this questionnaire in January 2005. After the flooding tragedy in September 2005, this community's perception of the importance of government support and public insurance might have changed depending on government response following the disaster.

(enlisting more of its scarce production resources) or contract debt for the amount of the net loss merely to maintain the level of imported goods purchased in the past. Production reorientation seems to be the most effective economic response to adapt to negative terms of trade within the agricultural sector. This means both diversifying crops to reduce the probability of being negatively affected by sudden price drops in the mono-crop, and moving to more profitable crops - crops whose selling prices are relatively higher, with a more stable demand and suitable to environmental and climatic conditions in the region. In this regard, Gay et al. (2004) analyse the high vulnerability in a coffee production region in southern Mexico to the extreme climatic conditions which are increasingly affecting the region. They warn of the low viability of growing coffee in these regions once the negative effects of policy changes and market instability have been internalised in this sector.

In a face-to-face interview conducted as part of the present study, Antonio Ruiz, Deputy Minister of Agriculture of Mexico, emphasized the need for differentiating between drought as natural disaster and persistent drought as part of the ongoing desertification process taking place in some regions of the country. In his view, it is too onerous for the government to keep paying indemnities or subsidizing extremely high insurance premiums in regions unsuitable for agriculture. Traditional crops such as grains have experienced a clear deterioration of terms of trade relative to garden vegetables over the past two decades in Mexico (ECLAC 2006). It is perhaps for this reason that 58% of interviewed farmers have plans to diversify to higher profit crops. Over 87% of interview subjects grow maize and coffee, crops whose prices have dropped dramatically over the past ten years. Of farmers who plan to stop raising a particular crop, 85% plan to phase out coffee and maize, especially in Cacahoatán, which is located in the heart of the coffee region of Chiapas. This is particularly important in light of the total elimination of tariffs on agricultural imports from Canada and the USA scheduled to take place by the end of 2008 in accordance with NAFTA. Fruits and vegetables would become more profitable to the farmers in this region given favourable climatic conditions and relative prices. In a World Bank report, Lederman et al (2003) present evidence suggesting fruits and vegetables to have higher comparative and competitive advantages to the Mexican agricultural sector, especially for export to North America under NAFTA. Even when freight and insurance costs are greater in the south for exporting to the USA and Canada, the relative higher availability of water in the south might make it still very profitable. However, as stated by most farmers and local leaders of farmers' unions in the interviews, lack of coordination between farmers and authorities impedes farmers' ability to receive accurate information and to obtain the knowledge required to change to these crops and, further, on how to afford rentable transport for their exports.

Encouraging crop diversification might improve the prospects of farmers and could help increase rural income in the region, lowering emigration. As shown in Fig. 6.3, the responses to the questions *"Do you have plans to diversify?"*, and *"Do you have plans to emigrate?"* present clear differences between the communities surveyed. Despite being the poorest community of the three surveyed, Escuintla has the highest proportion of farmers who plan to change crops (77%), and they are the least prone to

Fig. 6.3 Diversification and
emigration plans

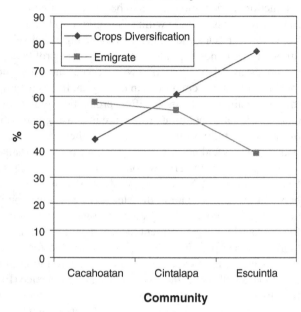

**Farmers' plans to diversify and emigrate by
community**

SOURCE: Saldaña-Zorrilla 2008.

emigrate (38% compared to the mean of 52%). In contrast, farmers in Cacahoatán
have the lowest propensity to diversify crops (44%), and are the most prone to emigrate (58%) despite this being the least marginalized community of the sample.

6.4.1.2   Emigration patterns in Mexico

As suggested by the results reported above, migration seems to be a question not just
of poverty, but of expectations of future income. People in communities repeatedly
damaged by natural disasters but without proper disaster risk management and
favourable entitlements have more pessimistic expectations of future income. In
turn, they more readily consider emigration as a coping strategy. These communities
are even more pessimistic than poorer communities. The presence of informal "travel
agencies" which arrange transportation for emigrants to Tijuana (the 3,500-km dis-
tant gateway across the border to the USA) was stronger in Cacahoatán, the "richer"
surveyed community. These findings provide empirical evidence for Todaro's postulates on emigration (2000). According to Todaro, migration is primarily an economic
phenomenon, which for the individual migrant can be a rational decision despite the
existence of urban unemployment. It follows that migration occurs in response to

rural-urban differences in expected income rather than actual earnings. The decision to emigrate is made to maximize expected lifetime earnings. For a given time period either the urban sector or jobs abroad appear more favourable. In other words, if the coping and adaptive capacity of vulnerable economic agents is not sufficient, they find emigration to higher income regions to be their best option. However, immigrants from the countryside have to face the prevailing precarious employment conditions once in the cities because of the current low capacity of the Mexican urban economy to absorb additional workforce. As migrants from rural areas do not get employed in urban industrial activities, they are forced to engage in low-productivity tertiary activities. In addition, a large portion of this underemployed population becomes an obstacle to an efficient allocation of public resources since the society is forced to provide large amounts of urban social infrastructure at the expense of directly public investments (Colosio-Murrieta 1979). Hard living conditions in urban areas provide pressure to emigrate abroad as well. About one-quarter of Mexico's population rely on some kind of agricultural activity (over 25 million people), and as conditions in the countryside continue to worsen and the urban Mexican economy continues growing slowly, rural workers will have no incentives to stop migration to the USA, where currently ca. 12 million of illegal Mexican migrants reside (US Census Bureau 2002). Although it might imply increasing remittances to Mexico, there are dramatic humanitarian consequences for migrants both to illegally cross the border as well as for subsisting illegally in a foreign country.

At the national level, emigration from Mexico has increased over the past two decades. Official figures report that whereas in 1990 0.24% of the Mexican population was residing abroad, by 2000 this figure had risen to 0.41% (INEGI 2005). Currently one out of three immigrants to the USA was born in Mexico (ca. 12 million) - the largest national group of immigrants in the country (US Census Bureau 2002).

6.4.1.3 Emigration in the surveyed communities

The proportion of Mexican population from the municipality of Cacahoatán living abroad (0.53%) is higher than the 2000 national average, while the municipality of Escuintla has a lower proportion (0.36%), though both figures exceed the Chiapas state average of 0.15%. Total emigration (domestic plus international) in Cacahoatán and Escuintla is 1.48% and 1.34% respectively.[6] This survey provides evidence in support of emigration as an important coping strategy by people in these communities. Natural disasters have motivated emigration, and farmers' plans to emigrate are influenced by the presence of relatives living abroad. The questionnaires indicated that this influence is, in fact, greater if these relatives emigrated due to disasters. Most importantly, recurring natural disasters in the past ten years, combined with

---

[6] Author's estimates using data from *Sistema Municipal de Bases de Datos*, INEGI (www.inegi. gob.mx). It should also be taken into consideration that official records tend to miss migratory moves taking place in two stages, that is, migrants who first move to cities within the same country and later leave the country.

Table 6.1 Correlations between plans to emigrate, disasters, and relatives abroad

		PlansEmigr	RelatAbroad	RelativsEmigrDis	MNatdis3
PlansEmigr	Pearson Correlation	1	.321(**)	.341(**)	.077
	Sig. (2-tailed)		.001	.001	.426
	N	123	112	91	110
RelatAbroad	Pearson Correlation	.321(**)	1	.378(**)	.051
	Sig. (2-tailed)	.001		.000	.600
	N	112	122	93	109
RelativsEmigrDis	Pearson Correlation	.341(**)	.378(**)	1	.251(*)
	Sig. (2-tailed)	.001	.000		.013
	N	91	93	105	98
MNatdis3	Pearson Correlation	.077	.051	.251(*)	1
	Sig. (2-tailed)	.426	.600	.013	
	N	110	109	98	133

the presence of relatives who emigrated due to disasters in the past, were shown to increase the desire of farmers to emigrate.

Of all the interview subjects, 41% had current plans to emigrate. As shown in Table 6.1, the correlation between people with "plans to emigrate" and those more frequently affected by natural disasters in the past 10 years –"No. of natural disasters"- is only 0.077 (row 1, column 4). However, the correlation is statistically significant for the variables "No. of natural disasters" and "relatives migrated due to (natural) disasters." The correlation with these variables is 0.251, which is significant at the 0.05 level (2-tailed test). The correlation coefficient between the variables "plans to emigrate" and "relatives living abroad" is 0.321, significant at the 0.01 level (2-tailed test) under Pearson's correlation test. However, the variable "plans to emigrate" shows higher values of correlation with "relatives migrated due to natural disasters".

The latter variable represents the existence of relatives who have used emigration as a coping strategy after losing assets and/or productivity due to natural disasters. The value of the correlation test for these variables is 0.341, the strongest observed. It also confirms the importance of networks of relatives in facilitating emigration, which is catalysed by the negative impact of natural disasters in the region.

Of the interviewees who expressed plans to emigrate, some 40% plan to emigrate to the United States of America (USA) - 15% of the total sample. One third of those interviewed would recommend emigration to a relative given current conditions. 38% of interviewees have relatives living abroad, of which 96% live in the USA (36%

6.6.2   Ignorance about risk prevention and paternalism

Authorities consider that insurance makes sense only if the disaster risk is reduced through appropriate public works which help to keep premiums down. Private crop insurance is barely utilized in the region, being contracted only by transnational companies such as Del Monte, Chiquita, and La Moderna. *Fondos* is an instrument unused by middle and low income farmers in this region. In fact, it seems to suffer from a lack of coordination between levels of government. Subsistence farmer leaders interviewed appeared to regard crop insurance with some scepticism, and in general they showed more confidence in construction projects for disaster mitigation and credit.

Overall, there is a widespread opinion among government authorities that the failure of public intervention in reducing farmers' vulnerability to market and disaster shocks is due to the passive attitude among the farmers themselves. It is felt that the farmers rely too much on a paternalistic government and show almost no initiative towards reducing their vulnerable conditions themselves. Farmers' leaders say that it is difficult to come up with initiatives given current low levels of education and the existing adverse endowment system. The former impedes them in devising technical solutions and the latter poses an obstacle to progress in their initiatives. Some government officials also reported local leaders' lack of interest in promoting vulnerability reduction measures whether undertaken by farmers or the government unless it brought them some personal financial benefit. Despite these differences, there is common agreement about the need for more and better institutional support in three respects: (*i*) financial support, (*ii*) production re-orientation, and (*iii*) replacing paternalistic assistance by investments in infrastructure.

6.7   Comparing resilience building 2004-2014

Researchers seldom come back to the surveyed areas. On December 2014, the author visited these three communities again in order to compare vulnerability and resilience conditions after ten years. In 2014, it applied 84 questionnaires to small-scale farmers and 21 interviews with local leaders and authorities. One important component of the 2004-2005 research consists of the hand delivery of a set of suggestions to reduce vulnerability to the government and leaders at the federal, provincial and local levels. In 2005, the author even talked personally to a number of public officers in charge of governmental programs for disaster risk reduction requesting their intervention on these communities based on this research findings'. They welcomed and acknowledged this research contribution to their decision making. However, such an intervention did not take place in practice. To the contrary, ten years after, the vulnerability conditions throughout these regions seem to be aggravated.

In October 2005 (nine months after the first survey took place), the surveyed area was again devastated. Hurricane Stan hit the coast of the state of Chiapas. It led to a loss of around 20% of the housing of the corresponding settlements, according to official loss assessments (CENAPRED-ECLAC 2006) as well as verified through this research field work in 2014.

The enormous devastation from hurricane Stan throughout the surveyed area led to a corresponding rebuilding and rehabilitation of the public and the private sector. Under constant accusations of public resources misuse and fraud from the Chiapas state governor, the public infrastructure was rebuilt and with governmental support, the private sector coped relatively well with the losses, as confirmed by interviews and questionnaires applied to the affected population.

Derived from the 2014 survey, it identified a relatively new source of risk arising from the environmental management in the region, with respect to some practices of the open-air mining industry during the last ten years. According to a recent research project carried on by the University of Sciences and Arts of Chiapas (UNICACH) and the IMTA (Spanish acronym for Mexican Institute of Water Technologies, a public sector research institute), during the last ten years the number of open-air mining exploitation concessions around natural protected areas increased dramatically, even though the current environmental legislation forbids such activities over there. Most of these mines are located on the mountains of the Sierra Madre de Chiapas. As these mines were dug on high slopes of the mountains, even normal levels of recent precipitation have dramatically eroded the region. If a hurricane with a similar intensity to Stan in 2005 occurred today, the negative impact would certainly be much greater on the surveyed communities (Fig. 6.7).

The conditions warn of the urgent need to:

- Ban all open-air mining activity in the region.
- Reforest with native species the eroded areas and restore soils in order to prevent dramatic floods over these communities again.
- Create living barriers with native species along the banks of rivers flowing at the edge of human settlements in order to smooth river overflows.
- Implement civil protection committees in every single community within the risk area.
- Conduct cost-benefit analyses of further disaster risk reduction measures in the area and implement the resulting optimal public works as soon as possible.

Concerning the population's perceptions, they have become overall more pessimistic about their future income expectations, given the disaster risk between 2004 and 2014. For instance, in the community of Cintalapa, there was a 23% increase in the number of respondents with plans to migrate out of the community in 2014 compared to 2004; it remains a null per cent of crop insurance coverage in both years and still practically none export.

6.7.1   Corruption

Unlike in 2004, the word "corruption" was frequently mentioned as their main obstacle to improve their life conditions. Corruption was also heavily pointed out as the main reason for the increasing number of mining concessions in the region as well,

Nueva Francia mine, June 2012, Sierra Madre de Chiapas, municipality of Escuintla. Photo by courtesy of *La Jornada.*

Nueva Francia mine, March 2013, Sierra Madre de Chiapas, municipality of Escuintla. Photo by courtesy of *Romeo García Cortés.*

Fig. 6.7 Fast-growing ecosystems devastation and erosion from mining companies in the Sierra Madre de Chiapas surrounding Escuintla 2012-2013

linking the state and federal governments with multinational mining companies. PROFEPA, the Mexican environmental protection agency, seems to be withdrawing from intervening in regard to these mines, warns Mr. Juan Carlos Castro, biologist and director of the natural protected area El Triunfo, which is a 119 thousand hectares rainforest located around these mines on the Sierra Madre de Chiapas. As part of his interview during this 2014 fieldwork, Mr. Castro also points out the need for a prompt reforestation and soils restoration instead of granting further mining concessions on these areas. For its part, Mr. Romeo García Cortés, Secretary General of *Escuintla Nuestro*, a local NGO with a long-standing resistance position against the ecosystems devastation from mining companies in this region, warns of the passivity from the local, state and federal authorities before claims from the civil society, including his NGO. He complains as well of anonymous death threats asking him to stop his denunciations.

6.8 Conclusions

There is general agreement among the stakeholders interviewed that the negative impacts of natural disasters are increasing and that the prevailing conditions of poverty and current economic policy are amplifying the negative consequences resulting from weather-related events. In the farmers' views, any solution must be accompanied by a comprehensive coordinated financial, economic and disaster management strategy in the countryside. They recognize that the complex economic-climatic hazards and stressors they face require more than social programs and post-disaster aid. Solutions must encompass production reorientation and disaster prevention instruments. There was widespread feeling among interview respondents that subsistence farmers in this region lack an effective long-term sustainable strategy to strengthen their adaptive and coping capacity. Despite their careful, innovative design, public instruments to deal with disaster losses (e.g. FOPREDEN) lack penetration. The prevailing conditions of marginalisation in this region, along with insufficient government investment in infrastructure, low use of credit, insufficient subsidies to crop insurance and a lack of investment in more profitable crops greatly reduces the communities' coping capacity when hazards strike, which in turn triggers migration.

Farmers' preferred sources of financing to cope with disasters in the region were found to be assistance from relatives and neighbours, and temporary government aid. Pre-disaster financing instruments, such as communal savings and insurance have been practically unused by these farmers. The low implementation of public disaster prevention instruments in the surveyed region can be explained in part by insufficient promotion by the federation and state governments, but more importantly, by the scarce risk analyses carried out by the state government. Given their increasing losses from disasters, farmers' expectations of improving their future living conditions are becoming increasingly undermined. Questionnaire responses showed that the farmers most determined to emigrate are not necessarily the poorest ones, but those with the most pessimistic expectations of future increases in their income. As vulnerability to disasters reduces farmers' expectations of future agricultural income, it increases their desire to emigrate. The lack of mechanisms to diversify crops appears to play a crucial role, as emigration is the preferred response of those farmers who are relatively less likely to diversify their crops. Due to the high number of potential migrants, an eventual free labor mobility agreement between Mexico and the USA might increase labor supply dramatically in the USA. However, sealing the border to further Mexican immigration may become hard to manage if one considers the intense trade between these two countries. Nevertheless, a cohesion fund to support Mexican rural economy may certainly contribute to smooth such a pressure. Most importantly, Mexico needs to generate better paid jobs at home by means of, first, enhancing productivity and commercialization conditions in the rural economy and, second, investing in trade infrastructure, i.e. maritime transport facilities (Saldaña-Zorrilla 2003).

Finally, there is a widespread demand from farmers for authorities in the region to be more actively involved in promoting vulnerability reduction measures. Government intervention should be more coordinated and anticipate negative outcomes resulting from uncertain trade conditions and increased incidence of weather-related disasters. This intervention should contribute to strengthening farmers' adaptive and coping capacity through mechanisms aimed at increasing their profitability and facilitating favourable endowments. Authorities, in turn, would like more initiatives from the farm leaders. These initiatives should be based on a realistic, inclusive evaluation of communal needs which goes beyond personal opportunism. Both farmers and authorities require a better understanding of local needs to optimise the allocation of public resources to reduce vulnerability to hazards and increase the profitability of farm assets. All agree that this would improve farm incomes and significantly reduce poverty and out-migration in this region of Mexico.

Chapter References

Benson C, Clay E (2002) Vulnerability and the Global Economy. In: Kreimer A, Arnold M, Carlin A (eds) The Future of Disaster Risk: Building Safer Cities, Conference Papers. The World Bank, Washington, DC

CENAPRED (2001) Diagnóstico de peligros e identificación de riesgos de desastres en México: Atlas Nacional de Riesgos de la República Mexicana. Centro Nacional de Prevención de Desastres. Ministry of the Interior of Mexico, Mexico City

CENAPRED-ECLAC (2006) Características e Impacto Socioeconómico de los Huracanes "Stan"y "Wilma" en la República Mexicana en 2005. The Ministry of the Interior of Mexico in cooperation with ECLAC. Mexico City

Colosio-Murrieta, LD (1979) Urbanization and Economic Development in Mexico. International Institute for Applied Systems Analysis (IIASA), Working Paper. Laxenburg, Austria

CONAPO (2004) Indice de Marginacion 2000: por Entidad Federativa y Municipio. Consejo Nacional de Poblacion. http://www.conapo.gob.mx/00cifras/2000.htm Accessed 15 May 2004

Conde C, Gay C (1999) Impacts of Climate Change and Climate Variability in Mexico. Acclimations, Newsletter of the US National Assessment of the Potential Consequences of Climate Variability and Change

Eakin H (2003) The Social Vulnerability of Irrigated Vegetable Farming Households in Central Puebla. The Journal of Environment & Development, Vol. 12, No. 4, 414-429 (2003) DOI: 10.1177/1070496503257733. SAGE Publications

ECLAC (2006) México: Crecimiento Agropecuario, TLCAN, Capital Humano y Gestión del Riesgo. United Nations Economic Commission for Latin America and the Caribbean, Mexico City

ECLAC (2001) Instituciones y pobreza rurales en México y Centroamérica. United Nations Economic Commission for Latin America and the Caribbean, Mexico City, p 30

Emanuel R, Greenberg J (2000 Lluvia Enojada-Tyoo Kuasi': ThePolitical Ecology of Forest Extraction in the Sierra Chatina, Oaxaca, Mexico. In Journal of Political Ecology:Case Studies in History and Society. Vol. 7. Journal of the Political Ecology Society. Arizona

Figueroa A (1991) Desarrollo agrícola en la América Latina. In: El Desarrollo desde adentro. El Trimestre Económico No. 71. ECLAC. Fondo de Cultura Económica. Mexico City

Gay C, Estrada F, Conde C at al (2004) Impactos Potenciales del Cambio Climatico en la Agricultura: Escenarios de Producción de Café para el 2050 en Veracruz (Mexico). El Clima, entre el Mar y la Montaña. D. Rasilla Álvarez (comp.). Universidad de Cantabria: 651–660. Santander

Guy Carpenter (2006) Tropical Cyclone Review 2005, Instrat Briefing http://gcportal.guycarp. com/portal/extranet/popup/pdf/GCBriefings/Tropical_Cyclone_Review_2005.pdf   Accessed 12 January 2006

Hernández-Trujillo JM (1997) Evaluación y perspectivas de los fondos de aseguramiento agropecuario (Assessment and perspectives about insurance crop Funds). Departamento de Economía. UAM-Azcapotzalco, Mexico City

INEGI (2005) National Institute of Statistics, Geography, and Informatics, http://www.inegi.gob. mx Accessed 5 December 2005

Krugman P, Obstfeld M (1997) International Economics: Theory and Policy. Addison-Wesley, Reading, MA

La Red (2002) Desinventar y los registros sobre escasez de agua en México: 1970-2000. In: García, V, De la Parra M, Nueva época. CIESAS, Mexico City

Lederman D et al (2003) Lessons learned from NAFTA. In: The World Bank Group, NAFTA's remaining trade barriers. Washington, DC

Linerooth-Bayer J, Vari A (2004) A Model-based Stakeholder Approach for Designing Disaster Insurance. In: Demeter K, Erkan NE, Guner A (eds) The Role of Local Governments in Reducing the Risk of Disasters. The World Bank Institute, Washington, DC, pp 159–185

Montserrat H, Chavez M (2003) Tres Modelos de Política Económica en México durante los últimos sesenta años. Análisis Económico No. 37., Vol. XVIII. Universidad Autónoma Metropolitana. Mexico City

Munich Re (2005). Net Cat Service http://www.munichre.com/ Accessed 7 November 2005

Norris FH, Stevens SP, Pfefferbaum B et al (2008) Community resilience as a Metaphor, Theory, Set of Capacities and Strategy for Disasters Readyness. American Journal of Community Psychology 41:127–150

Saldaña-Zorrilla, SO (2003) El Comercio México-Unión Europea (The Mexico-European Union Trade). Análisis Económico No. 37., Vol. XVIII. 1er. Semestre 2003. Universidad Autónoma Metropolitana-Azcapotzalco. Mexico City

SEDESOL (2005) Pobreza y desigualdad a nivel municipal y estatal. Subsecretaría de Prospectiva, Planeación y Evaluación, Mexico City

SIACON (2005). Sistema de Información Agropecuaria de Consulta. Ministry of Agriculture and Livestock.   http://www.siap.sagarpa.gob.mx/sistemas/siacon/SIACON.html   Accessed   5 December 2005

Todaro MP (2000) Urbanization and Rural-Urban Migration: Theory and Practice. Economic Development. Seventh Edition. Addison Wesley Longman. Essex

US Census Bureau (2002). http://www.census.gov/acs/www/ Accessed 5 May 2002

World Bank (2004) Poverty in Mexico: an assessment of conditions, trends, and government strategy. Report No: 28612. Mexico City

World Bank (2002) Country assistance strategy of the World Bank Group for the Mexican United States, Washington, DC

Wodon Q, Velez E (2000) Poverty and Inequality. In: The World Bank Group, Mexico: A Comprehensive Development Agenda for the New Era. Washington, DC

Yúnez-Naude A, Barceinas F (2003) Los Impactos del TLCAN (NAFTA) en el sector agrícola de México. Mimeographed. Department of Economics, El Colegio de México, México City

Chapter 7
Conclusions

Vulnerability to natural disasters and market uncertainty in Mexico is significantly contributing to the enlargement of poverty and inequity. The current hazards management implementation is insufficient to cope with their negative outcomes, and the unfavorable productive and entitlement structures contribute to their amplification in particular geographic regions and economic sectors. Despite its novel design, the lack of penetration of disaster prevention instruments is endemic among the poor, and persists in rural areas. Although natural disasters and current trade conditions affect mainly the agricultural sector in this country, their negative impacts spill over to other regions and sectors. It justifies a more active public intervention that in turn reduces inequity.

The analysis conducted throughout the present book proves that the number of people living in poor conditions is aggravated by disastrous losses and dropping prices of the regional commodity. It, consequently, warns of the risk of an even larger percentage of the population falling below the poverty line in the future, given the current trend of increasing frequency and economic losses from natural disasters, as well as the decreasing agricultural prices in the country.

This underestimation of the negative impact of climatic conditions is evidenced in the current social and productive expenditure distribution, which tends to ignore the protection of *ex-ante* assets of the most exposed population. Triggered by a vulnerable agriculture, the rural economy started a decomposition process during the past two decades, which has consequently stimulated out- migration to large cities and abroad. In order to counteract that process, this book suggests not only expanding the coverage of disaster prevention measures, but simultaneously expanding productive infrastructure to boost productivity along with more favorable production projections (considering the relative price structure), and expanding credit access and risk-sharing mechanisms.

The liberalizing economic policy of the last three decades has largely contributed to increasing that vulnerability, as described in Chapters 2 and 3. Lack of investments to expand irrigation and further physical infrastructure, rural credit reduction,

© The Author 2015   113
S.O. Saldaña-Zorrilla, *Natural Disasters, Foreign Trade and Agriculture in Mexico*,
SpringerBriefs in Environmental Science, DOI 10.1007/978-3-319-17359-7_7

and the reduced share of insured farmers seem to counteract by far the current aid to the poor. In addition, the governmental withdrawal of support for the countryside with infrastructure and commercialization facilities, may not only have increased poverty in the country (as there have not been realistic options to replace its intervention), but also indicated a certain naïveté on the government's part. In comparison, Mexico's trade counterparts continue to support their own countryside, both in the form of facilities and subsidies, increasing the trade deficit in that sector, and thus joining together to drop agricultural prices and rural incomes in turn.

The models presented in Chapter 5 confirm this book's hypotheses. Additionally, this modeling supports the decision-making process in issues concerning poverty, migration and disasters' vulnerability by levels of geographic desegregation for the whole country. It analyzes the effect of relevant determinants of income: productive infrastructure, price structure, financing instruments, asset distribution and external shocks. Estimating the differential influence of hazards on the different municipalities of Mexico allows for a more accurate and impact-based public intervention. Forecasting the return of investments in a given municipality together with those in surrounding municipalities enables the setting of priorities based on the goal variables of the national development plan. The desegregation of the model allows analyzing the differential impact of a variable depending on the marginalization degree and location. It enables accounting for disparities between crucial dichotomies of the country: north-south, rich-poor, and rural-urban. As observed, the negative impact of the analyzed hazards is higher among particular regions and sectors. It can be counteracted by strengthening households' coping and adaptive capacity based on moving to a more favorable productive structure and improving the access to financial, physical and human assets. Most likely due to the current entitlements system, better-off economic agents do not filter benefits to the others. It calls for additional and more selective public investments to close the increasing inequity gaps according to the corresponding poverty determinants.

In general, a marginalized population tends to be more vulnerable to weather-related disasters, whereas a non-marginalized population is somewhat more affected by economic stressors. The higher reliance on self-consumption among marginalized populations seems to explain that relation. Their lesser involvement in the monetary economy makes their subsistence more vulnerable to climatic conditions, whereas adverse productive and trade conditions are an obstacle from enrolling into the commercial agriculture, as they are unable to obtain production exceeding their basic needs on a regular basis. It impedes them from accumulating assets significantly to overcome poverty. Investments in instruments for reducing disaster loss possess comparatively higher benefits in reducing poverty across most marginalized regions, whereas enhancing intra-branch terms of trade and lowering transport costs brings greater benefits to those less marginalized. A public intervention biased to boost the commercial agriculture and reduce disaster loss in Mexico may limit poverty enlargement and sustain the domestic food supply. But given the spillover effect of poverty in Mexico, it will not be translated into permanent sustainable poverty reduction unless combined with substantial achievements in protecting and accumulating assets of subsistence farmers and the poor in general.

The most significantly migration prone population segments in Mexico are those with considerable income decreases and a large frequency of disasters in the last ten years. In addition, they have comparatively higher education levels and arise from marginalized isolated regions. In absence of a promising welfare horizon, the benefits of education in remote areas are frequently wasted given the lack of opportunities to make use of such a human capital, subsequently drained through emigration. Public investments in productive infrastructure, financial support, and projecting to build a more favorable trade environment in these regions may counteract that trend –and in turn, reduce social pressure in urban areas and impasses at the Mexico-USA border. Networks of migrant traders prove to be crucial for labor mobility, evidenced by the usual remote origin of migrants. As long as transport infrastructure does not become an effective input to boost trade and productivity in rural areas, it will continue serving just as infrastructure for migrants trading.

One of the main obstacles in carrying out public works to mitigate natural disasters losses is the current failed budgetary decentralization operation, as derived from the consultation with policymakers and authorities presented in Chapter 6. As it is mandatory for municipalities to identify their own disasters' risk in order to apply for the corresponding federal fund, the incapacity of most of them in fulfilling that requirement leaves most of these resources unused. It follows that the obtained loans and funds aimed at disasters' prevention are used mostly to finance current expenditure.

The empirical evidence from the stakeholders' interviews presented in Chapter 6 shows that the complex economic-climatic process behind poverty is sharpened by pernicious political practices at the three administrative levels. The insufficient promotion of strategies from the federation and state governments fits with the asymmetric capacity among municipal authorities and results in a lack of action on the part of the authorities. The disaster reduction and productive re-orientation strategies are often unknown to local leaders and authorities in regions that are poor and frequently affected by natural disasters. In light of these facts, the preferred coping strategies of the affected community are often assistance from relatives and neighbors, complemented with governmental aid. After the initial shock of the disaster, no preventive action is mostly taken for future events. The lack of planning for future events creates a vicious circle of economic vulnerability which pushes this population into an increasing state of poverty.

The withdrawal of the government from supporting agricultural commercialization has not led to the expected emergence of efficient private market instruments, but to the proliferation of speculators and a number of trade intermediaries. Such conditions dissolve any possible benefit for farmers when agricultural prices soar. Something similar occurs with rural credit, currently commanded by informal loaners at dramatically high interest rates and disadvantageous granting conditions, evidenced in the surveyed communities of this book. Communities more frequently affected by natural disasters and decreasing incomes during the past decade proved to be the most likely to view and implement out- migration as a coping strategy. Indeed, the communities more prone to emigrate are not the poorest ones of the sample. It suggests that the will to emigrate responds to negative expectations of future agricultural incomes rather than to poverty itself. However, the economic dynamism of the urban economy in the area is too modest to absorb them, forcing them to migrate farther. The dramatic proliferation of bus companies in the surveyed region, exclusively offering trips to the Mexico-USA border support this trend. Ironically, Mexican migrants work predominantly in the agricultural sector in the USA, contributing to an increase in productivity by lowering labor costs and in turn boosting USA agricultural exports to Mexico.

According to stakeholders' views, assets accumulation of individual households (both physical and human) along with building more favorable entitlements is crucial for reducing poverty in the region, and current social polarization in turn. More credit access and favorable market re-orientation based on crops diversification and rotation have been shown to be the most effective instruments to increase expectations among surveyed farmers and thus to lessen migration. Nevertheless, the benefit of insurance could not be conclusively tested given its widespread absence. In addition, the overall prevailing mutual distrust among stakeholders and the institutions –most likely derived from a recent past of weak democracy- seems to be impeding the implementation of the right programs.

During this research, the remarkable governmental concern about poverty and inequity in Mexico was obvious. However, in practice, the prevailing excessive belief in market instruments and confidence in budgetary decentralization seem to reduce the scope of the public intervention. It has opened the floor to a diversity of views concerning solutions to cope with hazards, often leading to implementing mutually annulling actions among public agencies at the three administrative levels or, perhaps worse, not implementing any. This is understandable if one considers the current lack of an integral vision concerning natural hazards and economic stressors from both the scientific, as well as the policy-making community. The findings from the present book regarding poverty and development confirm the need for further analysis addressing the coupled economy-nature relations based on a cross-section analysis. Mexico needs to actively implement multiple interconnected municipal projects (both rural and urban) for reducing vulnerability to natural and economic hazards coordinated by the social development authority. These should be carried out based on a transparent and stakeholders-inclusive mechanism so as to optimize economic resources and at the same time make sure that progress becomes compatible with equity.

Appendix A: The dynamic of economic vulnerability

Is high vulnerability responsible for low adaptability, or low adaptability the reason for high vulnerability? That question is the core of the discussion about how to assess the negative impacts of climatic events. It can be summed up in the following two positions: (1) the *vulnerability-adaptability:* identifying vulnerability in order to maneuver with adaptability measures; and (2) the *adaptability-vulnerability*, that is, studying adaptability to understand and further influence vulnerability. In this regard, supporters of the vulnerability-adaptability position (Kelly and Adger 2000, Chambers 1989) emphasize social vulnerability, and consider vulnerability to be a causal variable in this relation, where being vulnerable influences the ability of the systems units to cope with external stresses. In this perspective people are considered as risk managers subject to a relative context and restrictions. The adaptability-vulnerability position considers adaptability to be an independent variable explaining vulnerability, where vulnerability is a residual of adaptive capacity after a given climatic event. As in the first position, interdependence between adaptability and vulnerability is present here as well, but the causal force in this case is adaptability. However, which position is right depends rather on at what side of the dichotomy one locates assets and entitlements. As pointed out by Schjolden (2003) when comparing adaptability in Norway and Mozambique, vulnerability and poverty are not equal, and calls for an analysis of adaptability indicators to make them relative to economic and social stressors. She includes assets and entitlements in Vulnerability, falling into Position 1. However, given that adaptive capacity is rather a result of coping capacity, as seen in Section 2, we have more reasons to associate entitlements and assets with adaptability than with vulnerability itself in this paradigmatic cause-effect relation. if one considers entitlements and assets to be part of adaptability, expanding this concept to embrace coping capacity, one can asses vulnerability in the form expounded in this Appendix, making it easier to quantify vulnerability.

© The Author 2015
S.O. Saldaña-Zorrilla, *Natural Disasters, Foreign Trade and Agriculture in Mexico*,
SpringerBriefs in Environmental Science, DOI 10.1007/978-3-319-17359-7

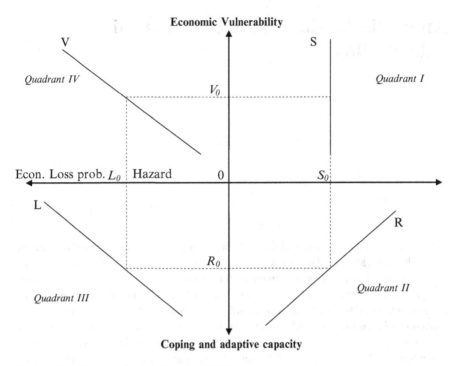

Fig. A.1 The dynamic of economic vulnerability

In this approach, as observed, the level of economic vulnerability to shocks is permanently interacting, and therefore can be modified by means of strengthening or weakening one or more of its constituting elements. Figure A.1 is a theoretical illustration of the dynamic of economic vulnerability. As one moves from the origin (intersect 0) values on the axis increase, that is, all values on the axis are positive.

The right x-axis of the graphic represents external shocks, which can be a perturbation or a stress in the form of a natural hazard, a sudden export restriction, etc. The upper y-axis represents the vulnerability level of a given unit of analysis. External shocks are independent of the vulnerability level of our unit of analysis, and so, a given external shock can be corresponded by different levels of vulnerability; Hence, this relation is represented by a vertical line; we term quadrant I the area depicted by these two axes.

The lower y-axis represents the variable *Coping and adaptive capacity*; it is a function of adaptive (risk management plus protection levels) and coping capacity (entitlements and assets). Coping and adaptive capacity–hence CA-are related with the variable hazards on the right horizontal axis shapes quadrant II. The slope of the respective curve, R, is determined by an entitlements and assets system; the higher the slope, the less favorable entitlements and assets system to our unit of analysis.

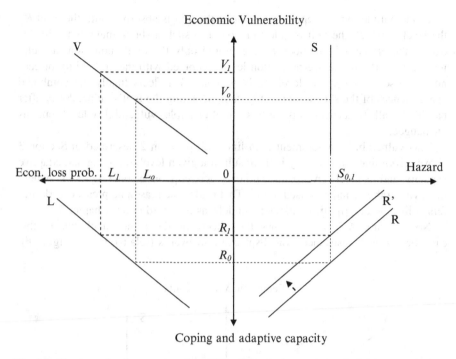

Fig. A.2 The dynamic of economic vulnerability after reducing coping and adaptive capacity

Curve R in quadrant II is the relation between unit's coping and adaptive capacity and hazards, which is a negative function: As hazards increase either severity or frequency, a lower CA level is related with them (Figure A.2).

The left x-axis is the *economic losses probability* inherent to our unit of analysis, and is a variable representing the likelihood of having a certain amount of economic losses as the consequence of a given shock affecting our unit, given prevailing coping and adaptive capacity. Both CA and the economic losses probability axes shape quadrant III. The relation between CA and probability of economic losses is negative, since as unit's CA decrease, the presence of economic damages is more likely to occur, represented by curve L.

Quadrant IV is depicted by the economic losses probability and axes. In the same way, economic losses probability contributes to increasing vulnerability, thus, the relation vulnerability-economic losses probability is positive.[1]

The graph above illustrates vulnerability relations with an initial equilibrium, where none of the components of vulnerability reduce or increase vulnerability level. So, V_0 represents initial level of vulnerability, S_0 is the shock occurrence at a

---

[1] In the last instance, the nature of vulnerability as a potential event forces our analysis to deal with, above all, probability terms.

given magnitude, whose repercussions upon the unit is absorbed until the level R_0, the level at which the unit is able to respond to such a shock magnitude. To this extent, the effects of the shock have explored only the unit's non-monetary elements. Now, the resistance-protection level combined with the probability of economic losses is set at the level L_0. Now, curve V reflects finally the combined consequences of the shock upon the system, depicting the unit's vulnerability after passing by all its determinants. In this initial example, vulnerability level remains unchanged.

This vulnerability assessment is in line with position 2 described in Section 3 above, given that vulnerability is the result of a given level of coping and adaptive capacity, and its level can be estimated only derived from a concrete shock –where most vulnerability indexes used to fail. That is also the reason he present work prefers holistic to deterministic disasters models, as observed in our Chapter 5.

Now, Figure A.3 below shows the effects of the same shock but in this case the curve R, resistance, was displaced downwards (relative to the origin, 0^2).

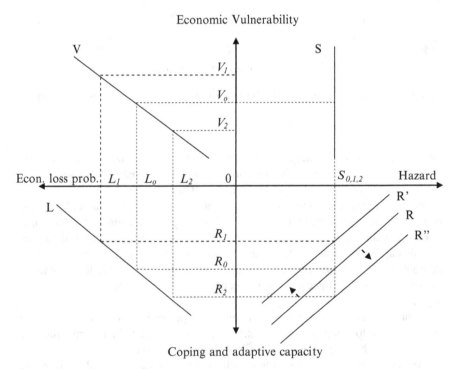

Fig. A.3 The dynamic of economic vulnerability after strengthening coping and adaptive capacity

---

[2] When we say downwards concerning curves displacement, we are using the reasoning of a typical x-y graph. Hence, saying downwards means only a move (displacement) of the curve approaching zero, the origin.

That displacement can be a result of failures in risk management in a given community, less access to credit for reconstruction, etc. The curve R' possesses lower CA values for all shocks compared to R. The same shock implies thus a lower resistance-protection level, moving from R_0 to R_1, corresponding to lower CA levels for all kinds of hazards. Given the relation where reductions in unit's CA increase the probability of economic losses (from L_0 to L_1, in this case), the resulting level of vulnerability increases from V_0 to V_1. The increase in vulnerability is thus the difference $V_1 - V_0$.

The same mechanism applies conversely for a displacement in the opposite direction of R. For instance, strengthening an element of the curve R leads to reducing vulnerability levels, other things being equal. The aim of the present work consists in identifying strategies that displace the curve R, to thus reduce the probability of economic losses to, in the last instance, reduce economic vulnerability. If curve R displaces from R to R", the external shock of Figure A.4 produces less economic losses than those derived from R and R', given that $L_2 < L_0, L_1$. It leads to reduction in vulnerability relative to that derived from the same shock with lesser coping and adaptive capacity, as the case of R and R', moving down economic vulnerability to V_2. In this perspective, vulnerability is determined by CA.

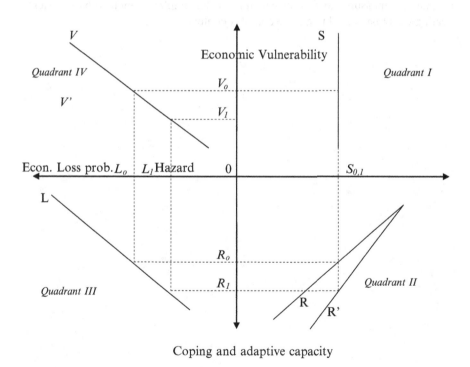

Coping and adaptive capacity

Fig. A.4 The dynamic of economic vulnerability after strengthening entitlements and assets system

Something similar can be done if one makes improvements to the entitlements and assets system, as for instance increasing units income, access to better infrastructure, less information asymmetry, etc. It would imply changing the slope of R from R to R', as seen in the graphic below. It reflects the fact that after improving assets and entitlements, the economic agent is better off to face a given hazard.[3] As observed, with the curve reducing its slope, the same hazard S_0 from Figure A.2, is smoothed by a higher coping and adaptive capacity level, which compared to the previous state increased from R_0 to R_1. For the given curve of economic losses distribution, it produces a reduction in economic losses, depicted by the reduction of probability of economic losses from L_0 to L_1. Now, this effect reflected on the vulnerability curve represents a reduction in vulnerability level equal to the difference $V_0 - V_1$. In that way, improving assets and entitlements to our unit of analysis implies strengthening its economic vulnerability for a given hazard.

As analyzed in this appendix, improving risk management contributes to reducing the probability of economic losses for a given hazard. However, risk management and protection systems have a limit, where additional progress cannot be reached despite increasing investments at a constant rate. For that reason, there is a need to combine adaptability with actions to improve assets and entitlements, that is, with coping capacity. In that sense, reducing vulnerability is not a mere result of adaptability measures to face hazard impact, but implies a more robust strategy dealing with that hazard in a more complex context.

---

[3] However, as the hazard increases, this capacity reduces approaching the previous levels, R.

Appendix B: Notes concerning construction of data bases

The natural disaster data used for this case study are reports of disasters from *LaRed* and *CENAPRED*, complemented by extrapolations based on records from other agencies. Data from these sources were integrated and harmonized as follows:

1. Reports of disasters were added from *CENAPRED* into the *LaRed* data base in some cases in order to scale down from state to municipal level data and in other cases to weight agricultural prices and losses. Data sampling was increased by 20%, providing valuable data on economic losses from disasters and numbers of hectares of crops damaged.
2. In order to make monetary data comparable, current economic losses were transformed to constant 2002 values.
3. In order to enlarge the sample size of economic losses, we included those observations without records of monetary loss but with number of hectares, by estimating monetary loss. Lost crop area was transformed into monetary losses by multiplying area by mean weighted price per ton ($/ton) and average yield (ton/ha). This increased the sample by 40%. Mean weighted agricultural prices were based on records of mean rural prices and production from SIACON (*Subsistema de Informacion Agricola*, SAGARPA 2004).

List of municipalities selected as main trading center cities

Acapulco de Juárez	Irapuato	Reynosa
Aguascalientes	Iztapalapa	Salina Cruz
Ahome	Juárez	Saltillo
Benito Juárez	León	San Luis Potosí
Cajeme	Manzanillo	Tampico
Campeche	Matamoros	Tapachula

(continued)

© The Author 2015
S.O. Saldaña-Zorrilla, *Natural Disasters, Foreign Trade and Agriculture in Mexico*,
SpringerBriefs in Environmental Science, DOI 10.1007/978-3-319-17359-7

(continued)

Celaya	Mazatlán	Tepic
Centro	Mérida	Tijuana
Chihuahua	Monterrey	Toluca
Coatzacoalcos	Morelia	Torreón
Cuernavaca	Nogales	Tuxtla Gutiérrez
Culiacán	Nuevo Laredo	Veracruz
Durango	Oaxaca de Juárez	Victoria
Guadalajara	Paz, La	Xalapa
Guaymas	Piedras Negras	Zacatecas
Hermosillo	Puebla	Zapotlán del Rey
Hidalgo del Parral	Querétaro	

Appendix C: How to introduce spatial data into the econometric notation?

The transformation of relative location information of contiguity into a weights matrix is the most employed way to relate the observations to each other in a network. Measures of contiguity rely on knowledge of the size and shape of the observational units such as municipality polygons depicted on a map. With that we can define which units are neighbours (have borders that touch) or represent observational units in reasonable proximity to each other using longitude and latitude.

For the sake of simplicity, assume that we are dealing with a geographic area comprised of five-regions, as shown in the figure C.1 below:

To construct the weights matrix, we record in each row of that matrix a set of contiguity relations associated with all regions, giving a value 1 to the w_{ij} element if region i borders to region j, 0 otherwise. All regions' connections may finally be represented in the following matrix:

$$W = \begin{pmatrix} 0 & 1 & 1 & 0 & 0 \\ 1 & 0 & 0 & 0 & 0 \\ 1 & 0 & 0 & 1 & 1 \\ 0 & 0 & 1 & 0 & 1 \\ 0 & 0 & 1 & 1 & 0 \end{pmatrix}$$

(1)

For instance, the element in row 1, column 3 means that region 1 shares a border with region 3, as seen in Figure D.1, and so on for every element in the matrix W. This matrix is symmetric and always has zeros on the main diagonal – because one region has no border with itself. To start applying this matrix, it must first be

© The Author 2015
S.O. Saldaña-Zorrilla, *Natural Disasters, Foreign Trade and Agriculture in Mexico*,
SpringerBriefs in Environmental Science, DOI 10.1007/978-3-319-17359-7

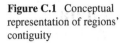

Figure C.1 Conceptual
representation of regions'
contiguity

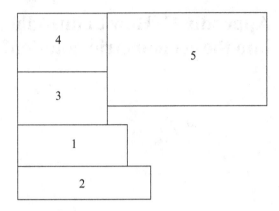

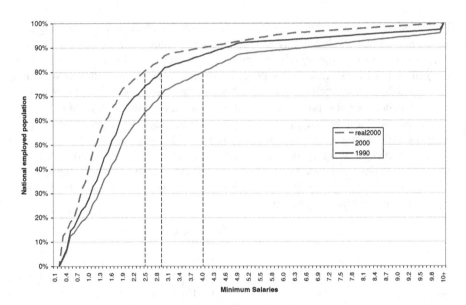

Fig. D.1 Distribution of workforce remunerations by minimum salaries (cumulative)

transformed to have row-sums of unity, conventionally called "standardized first-order" contiguity matrix (LeSage 1998), which can be denoted as C:

$$C = \begin{pmatrix} 0 & 1/2 & 1/2 & 0 & 0 \\ 1 & 0 & 0 & 0 & 0 \\ 1/3 & 0 & 0 & 1/3 & 1/3 \\ 0 & 0 & 1/2 & 0 & 1/2 \\ 0 & 0 & 1/2 & 1/2 & 0 \end{pmatrix}$$

(2)

We may now multiply C by a vector of observations of any variable -concerning the five regions of this simplified example- which we label y (5X5 matrix).[4] The product (Cy) represents a new variable, equal to the mean of observations from contiguous regions, which we denote as y*, expressed as follows:

$$
\begin{pmatrix} y_1^* \\ y_2^* \\ y_3^* \\ y_4^* \\ y_5^* \end{pmatrix} = \begin{pmatrix} 0 & 0.5 & 0.5 & 0 & 0 \\ 1 & 0 & 0 & 0 & 0 \\ 0.33 & 0 & 0 & 0.33 & 0.33 \\ 0 & 0 & 0.5 & 0 & 0.5 \\ 0 & 0 & 0.5 & 0.5 & 0 \end{pmatrix} \begin{pmatrix} y_1 \\ y_2 \\ y_3 \\ y_4 \\ y_5 \end{pmatrix}
$$

(3)

$$
\begin{pmatrix} y_1^* \\ y_2^* \\ y_3^* \\ y_4^* \\ y_5^* \end{pmatrix} = \begin{pmatrix} 0.5y_2 + 0.5y_3 \\ y_1 \\ 0.33y_1 + 0.33y_4 + 0.33y_5 \\ 0.5y_3 + 0.5y_5 \\ 0.5y_3 + 0.5y_4 \end{pmatrix}
$$

(4)

So,

$y_i = f(y_j), j \neq i$

Equation (4) may also be expressed in linear form in order to explain changes in y across space.

$$y = \rho Cy + \varepsilon$$

(5)

Where Here, ρ depicts a regression parameter and ε represents the stochastic error of the model. The parameter ρ shows the spatial spill over between neighbours in the data, reflecting the average influence of neighbouring on observations in the vector y. So, it enables us to estimate the share of total variation in y explained by spatial dependence (LeSage 1998). The Spatial Lag model as in equation (5) is an appropriate solution when spatial correlation appears among observations of the dependent variable.

Now, we can add explanatory variables to shape our model, representing it with the traditional matrix notation $X\beta$. So, the model equation (5) can be modified as follows:

$$y = \rho Cy + X\beta + \varepsilon$$

(6)

The inclusion of a spatial lag is equivalent to including an endogenous variable on the right_hand side of the equation in systems of simultaneous equations. This

---

[4] In our case of study model, both W and C matrices' size is 2,443 by 2,443.

would not be a problem if ρ was known, but most of the time ρ must be estimated simultaneously jointly with the other regression parameters. This means that estimation using OLS is inconsistent. The solution is estimation through maximum likelihood or the instrumental variables approach.

Another kind of spatial dependence occurs when the residuals are spatially correlated with the dependent variable. This kind of dependence occurs when values of adjacent observations move in the same direction because they have common or correlated unobservable variables. That is, lack of independence of the error between observations –as in temporal econometrics- would lead to inefficient estimates if ignored. The Spatial Error Model (Cliff and Ord 1972) as in equation (6) is established by means of a partition of the stochastic term into two parts, together with a spatially weighted matrix. In the case of spatial dependence present in the residuals we solve this problem by dividing the error term into two parts and incorporating the spatial error dependence in a similar way as before. The estimation method is maximum likelihood or generalized method of moments.

$$y = X\beta + \varepsilon$$

$$\varepsilon = \lambda C\varepsilon + \mu \tag{7}$$

In (7), λ is the spatial autoregressive parameter, C is the weights matrix, and $\mathbf{\mu}$ is a vector of i.i.d. errors with variance σ^2.

A first step to test for presence of spatial dependence is to make an exploratory data analysis to find indications of spatial autocorrelation, present in the data. We must thus test whether objects of similar values are more clustered than by mere coincidence. This is done using Moran's I test with a null hypothesis of no spatial clustering (Moran 1948, Cliff and Ord 1981, Upton and Fingleton 1985). The second step is to examine and solve for spatial dependence in the regression analysis. The classical estimation routine towards a proper model specification under the potential influence of spatial dependence is, for instance, Lagrange Multiplier tests, given in Florax et al. (2003). The initial model is estimated by means of OLS. The residuals are then used to test the hypothesis of no spatial dependence caused by an omitted spatial lag or by spatially autoregressive errors by use of two Lagrange Multiplier tests (the LM-lag test and the LM-error test), e.g., Anselin (1988) and Burridge (1980). When the hypothesis cannot be rejected (no spatial dependence is at hand) it means that the results from the OLS may be used. However, in the event that the hypothesis is rejected by both tests, a new model should be estimated. The proper model is indicated by the most significant LM test. In case that only the LM-lag test is significant, the next step would be to estimate a Spatial Lag Model, or, consequently, a Spatial Error Model if the opposite is indicated.

Appendix D: The official minimum salary in Mexico, a biased welfare indicator

The model presented in Chapter 5 makes use of an income-based measurement of vulnerability, whose link with poverty is addressed in this appendix. Further, it analyzes the evolution of real incomes (1 minimum salary) over the 1990s and describes its spatial distribution across Mexico in 2000. We show that apparent reductions in number of people earning below one minimum salary does indeed not reflect welfare improvements once the devaluation of one minimum salary over time has been considered.

Poverty

There exists no consensus about what issues embraces poverty. However, a point of agreement among the vast existing literature on this regard seems to be that poverty is a lack of opportunities to reach a basic minimum level of well-being. In turn, well-being can include several determinants, i.e., assets, infrastructure, social networks, institutions, human development, etc. This leads us to another point of agreement: poverty is a relative concept of "minimums of well-being" (as in SEDESOL 2002), "society's hopes and aspirations to be fulfilled" (World Bank 2001), or "basic needs" (Todaro 2000), which varies widely among and within societies. Other concepts of poverty go beyond tangible factors and include rights issues, relationships, powerlessness, socioeconomic exclusion, and even loss of dignity (i.e. Elankumaran et al 2000).

© The Author 2015
S.O. Saldaña-Zorrilla, *Natural Disasters, Foreign Trade and Agriculture in Mexico*,
SpringerBriefs in Environmental Science, DOI 10.1007/978-3-319-17359-7

Income and consumption criteria

Among measurements of poverty, there are two basic criteria: income and consumption. Income-based measurements of poverty embrace the total current monetary and non-monetary income a household obtains from different functional sources, including wages, salaries, dividends, rents, etc. Consumption-based criteria consist of the total amount of monetary and non-monetary expenditure a household makes in all possible items. The advantage of using consumption-based measurements of poverty relies on the fact that it can reflect smoothing strategies as a household implement in response to sudden income fluctuations. The use of income-based measurements is advantageous in that it can allow identifying income components whose strengthening can help to reducing vulnerability. Though the Ministry of Social Development at the Federal Mexican Government (SEDESOL) elaborates both measurements, only data of income measurements are available at municipality level.

For these reasons and in sight of the present work's aim, we make use of low-incomes criterion. As this measurement of poverty leaves out a household's assets, we include it in the model as an explaining variable within the variable 'marginalization'. The low-incomes line approached in this appendix is similar to the $1 per day standard criterion at Purchasing Power Parity (or *absolute PPP dollar* –Chan and Ravillion 2004). Given data availability, our low-incomes line is one official minimum salary, equivalent to average 2.3 US dollars during the period 1990-2000 (1990 constant prices), similar to the 2.1 exchange rate from the World Bank (1993 constant prices). This work's low-income line consists of incomes equal to or below 1 Minimum Salary (MS) by worker.

1. Minimum Salary across time and space

Before starting analyzing data, it is important to consider two facts. First, the aim of defining this low-income line in real terms is not to provide one more poverty measure of Mexico, but to allow us to identify spatially municipalities with low incomes and decreasing incomes in real terms. Whereas low incomes and poverty may keep similar geographic distribution, but changes in real incomes over the 1990s may diverge. In addition, this analysis does not account for the entire population living with less than one absolute PPP dollar, but with those workers earning less than that: consider that usually one worker still has to share his earned dollar with the rest of the family. Indeed, this analysis deals with just income and not with poverty itself. Secondly, a 30% real depreciation of Mexican general minimum salary (GMS) took place during the period 1990-2000 according to the International Labor Organization (ILO 2005). We had to modify the original municipal data by transforming minimum salary (MS) from nominal to real terms in order to obtain comparable low-incomes shares over a ten-year horizon. The latter provides a more realistic picture of the share of workers below the low-income line. Since the ILO estimations tend

Table D.1 Minimum Salary (MS) 1990-2000

	(1)	(2)	(3)	(4)	(5)
	Nominal MS (MXP/day)	Nominal Var. MS (year before)	Inflation rate (%)	Var. real MS (*) (%)	Real MS, MXP/ day, (constant 1990)
1990	9.14	16.66	24.62	-7.96	8.41
1991	10.79	18.03	25.02	-6.99	7.82
1992	12.08	12.03	16.73	-4.71	7.46
1993	13.06	8.08	10.45	-2.37	7.28
1994	13.97	6.97	7.12	-0.15	7.27
1995	15.85	13.42	24.02	-10.59	6.50
1996	19.55	23.35	41.42	-18.06	5.32
1997	24.30	24.33	23.34	0.98	5.38
1998	27.99	15.19	15.37	-0.19	5.37
1999	31.91	14.01	18.00	-4.00	5.15
2000	35.12	10.06	10.08	-0.02	5.15
Average	19.43	14.74	19.65	-4.91	6.46
Var. 1990/2000	284.29				-38.77
Sum		162.11	216.17	-54.05	71.10

Elaborated with data from *Comision Nacional de Salarios Minimos*, BANXICO and INEGI
(*) Discounting annual inflation rate.

to smooth the MS depreciation in Mexico to harmonize within Latin-American countries, we made our own estimations. According to this work's estimations, the real MS depreciation reaches 39% (see Table D.1 above).

Whereas in nominal terms accumulated growth of GMS is 162% between 2000 and 1990, in real terms (discounting 216% accumulated inflation rate) it dropped 54%. Real MS at constant prices 1990 decreases nearly 40%. One can observe the dramatic fall in real value of minimum salary in 1995 and 1996 -column (4)-, which coincides with the Mexican financial-economic crisis started in 1994. Though the federal government increases minimum salary, these increments do not always keep proportion with prices increments, which explains its depreciation in real terms.

In addition, we found different results in shares of labor force earning 1MS when taking minimum salary in nominal and real terms. In nominal terms, the share of labor force with income below 1MS decreased by 22% over this period (poverty reduction), but if one uses real minimum salary, the share of national population with incomes below 1MS increases 13% (poverty rise) –see Table D.2 below.

These estimations diverge from those of the World Bank's PovcalNet. According to this on-line calculation tool, the percentage of national population living below the poverty line in Mexico in 1990 was 17%. The estimation for year 2000 from PovcalNet drops to 10%, contrasting with our rise to 41%. Two remarks may explain this fact: first, here we are dealing with exclusively workforce, whereas PovcalNet's poverty line embraces total population. Secondly, given the increased labor market flexibility in Mexico since the end of the 80s (De la Garza 1990, De Buen 1989),

Table D.2 Changes in workers income below 1 Minimum Salary (1990/2000)		Workers income < 1 GMS
		(% of workforce)
	199013	28
	2000	22
	Real 2000	41
	Difference 2000/1990	-6
	Difference Real 2000/ 1990	

employed people receives each time lower remunerations from usually eventual jobs –paid either by worked hours or even by hour-fraction.[5] These changes in labor remunerations have been responding more to industries' workforce requirements, pressing for short-term –if any- contracts, as well as to smooth unemployment rise. The observed syndicates' weakening during this period was the key in achieving that (Perez 2002, Garabito 1995). One has also to account for the labor force growth over the 1990s, which has by far exceeded labor demand. As further explained, increases in jobs demand make pressure to reducing salaries, other things equal, in order to contain unemployment rise.[6]

2. Distribution of employed population by salary levels

Figure D.2 shows the aggregate picture of labor force by levels of salary. This graphic relates, on the x-axis, levels of salary with, on the y-axis, percentage of labor force. The curve presents accumulated values. The blue line represents incomes distribution in 1990, and the red line incomes distribution in 2000. Comparing these curves, one can observe an improvement, especially for low-salaries workers, as the curve displaces to the right. It reflects the fact that, for instance, in year 1990, 80% of national employed population was earning 2.9 MS, and for 2000 they achieved to earn up to 4.0 MS. In this perspective, these incomes distribution curves seem to have improved incomes for all cumulated percentiles of workers between 1990 and 2000 –in general measured by its displacement to the right. However, if one adjusts salaries to inflation (as in the previous section), this distribution changes substantially. The red dotted line represents real incomes dis-tribution in year 2000. Taking into account the ca. 40% depreciation of the mini-mum salary once adjusted for inflation, the leftwards displacement of the curve shows that if the minimum salary were adjusted for inflation, 80% of the national

---

[5] This kind of salary flexibility fits with the neoclassical approach of labor markets, where wages adjust as the marginal productivity of labor in a perfectly competitive economy. This fact raises arguments to associate reductions in the capital/labor ratio with low salaries, as further applied in the next section. See Krugman and Obstfeld 2006 and Sala-i-Martin 2000.

[6] Note that real salary depreciation is a plausible measure to employ exceeding labor force only in sectors with high elasticity of substitution labor-capital, further explained.

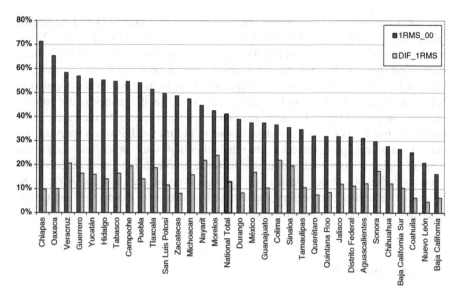

Fig. D.2 Share of low income workforce in 2000 and change 2000/1990

population has reduced incomes from 2.9 to 2.5 minimum salaries, and so on for other employed population shares. In addition, comparing earnings distribution curves of real 2000 and 1990 in Mexico, the real 2000 curve has a comparatively (to 1990) higher slope between 0 and 2 MS, representing a higher concentration of people below this threshold than before. Put in terms of a population pyramid of incomes, it is equal to an enlargement of the base.

3. Spatial distribution of low income

At national level, 41% of labor force earned less than 1 minimum salary in 2000, bringing down 13% of labor force from upper levels from 1990 to 2000. At state level, the 32 states of Mexico increased this share during the 1990s. Southern states register the highest shares of labor force earning less than 1 minimum salary (2 USD a day), remarkably high in Chiapas (71%), Oaxaca (65%), Veracruz (58%) and Guerrero (57%). The lowest percents correspond to states in the north of the country, namely those bordering the USA: Baja California (16%), Nuevo Leon (21%) and Coahuila (25%). From that fact, we deduce that this difference is influencing the northwards pattern of domestic emigration during the 90s in Mexico.[7]

---

[7]For an analysis on labor remunerations increase in the North of Mexico possibly linked with Maquiladoras in the framework of the North America Free Trade Agreement, see Lederman et al. (2003).

The states with highest *increases* in share of labor force earning below 1 minimum salary between 1990 and 2000 are Morelos (24%), Colima (22%), Nayarit (22%), and Veracruz (21%), located in the West and South of the country. Although these states have shares of low-income labor around the national mean, but this worsening along this period shows signs of vulnerability as putting additional population below the 2USD a day poverty line. Except for Veracruz, the change in the poorest Mexico's states -Chiapas, Oaxaca, and Guerrero- show moderate increases, most likely because further incomes reductions seem to turn smaller when approaching the bottom. For its part, Veracruz presents both a high share of low-incomes labor force (3rd. place) along with a dramatic increase in low income labor force shares over this period (4th. place) –see chart below. As observed, low incomes characterize historically marginalized regions, but the highest decreases in incomes are taking place along non-extreme poor regions.

4. Final comments

Roughly, the southern states register the highest shares of workforce with incomes below one minimum salary, whereas the north has the lowest. Besides, the highest per capita disasters losses regions, relative to insurance access, are located in the south, which may explain the northwards pattern of migration in Mexico. The real depreciation of the minimum salary of the 1990s was more noticeable in non-marginalized regions as they are more dependent on monetary incomes, but also because additional reductions in real incomes in marginalized regions imply small monetary variations. The highest increases in shares of workforce earning below 2 minimum salaries are in non-poor states, surely explained by their higher value of their assets. These results warn about the relevance of strengthening the implementation of adaptive instruments of the poor but also of the non-poor but highly exposed, unless one does not wish to enlarge poverty.

Appendix E: Data basis and series construction of the econometric model

The natural disasters data used for this case of study are reports of disasters from *LaRed* and *CENAPRED*, complemented with extrapolations based on registers from other agencies. The integration and harmonization of these two data sources was made as follows:

4. It added reports of disasters from *CENAPRED* to *LaRed* data, trying to down-scale state-level data to municipal level, when needed, by weighting with reference to particular criteria, depending on the case. It increased the data sampling in 20%, providing valuable data on mainly economic losses from disasters and damaged hectares of crops.
5. To make comparable monetary data, we transformed every observation regarding economic losses from disasters from current to year 2002 constant prices.
6. One filtered and left out observations regarding anthropogenic disasters, forest fires.
7. Observations at only state-level were distributed into municipalities by dividing state amounts by municipal population.
8. In order to enlarge the sample size of economic losses, those observations without registers of loss but with number of hectares were included by estimating their monetary loss. Area was transformed into monetary losses by multiplying it by mean weighted price per ton ($/ton) and further by average yield (ton/ha). The latter increased 40% the number of observation for the variable economic losses from disasters (L). Mean weighted agricultural prices were based on registers of mean rural prices and production from SIACON (*Subsistema de Informacion Agricola*, -SAGARPA 2004).
9. Hereafter, I filtered observations by decade, taking the 1970s, 1980s, 1990s, and the period 2000-2002 separately.
10. It estimated subtotals of each disasters-variable by municipality, so as to get uniform series by municipality along decades in cumulated values.

© The Author 2015
S.O. Saldaña-Zorrilla, *Natural Disasters, Foreign Trade and Agriculture in Mexico*,
SpringerBriefs in Environmental Science, DOI 10.1007/978-3-319-17359-7

11. It standardized data code based on the INEGI municipal code, to be used as *key* to join the data basis under the *GeoDa* and *Arc-View* softwares. So the software can identify and add variables columns to the final data basis. Also, I saved separately every decade as 'dbf' file, which enables it to be read by GIS software.

12. Shape file and variables tables are joint, ready to start model specification. Variables in the joint shp file include: share of labor by income classes, domestic, international and general emigration, population, irrigation, insurance, credit, marginalization index, etc (Tables E.1 and E.2).

Table E.1 Variables characteristics

VARIABLE	DATA CODE	PERIOD	UNIT	GEOGRAPHIC AGGREGATION	OBSERVATIONS
LnLIP - Low Incomes Population	2MS_00	2000	% of workforce, logarithm	Municipality	2,443
LnFA - Fixed Asstes	lnFA89p1	1989	Pesos per capita, logarithm	Municipality	2,443
LnD- Losses from Natural Disasters	lnLCS_P	1990-2000	Pesos per capita (constant 2000), logarithm	Municipality	2,443
ΔPA - Agricultural Prices Variation	Pa_Var	1990 / 2000	Pesos/ton (over 440 crops mean weight)	State (set to municipalities)	32 (2,443)
LnDIST -Distance to trading center	lnMind	1990	Meters, logarithm	Municipality	2,443
LnINS - Insurance	lnINS1	1990	Agricultural units with insurance (%),logarithm	Municipality	2,443
LnIRR - Irrigation	lnIRR1	1990	Agricultural units with irrigation (%),logarithm	Municipality	2,443
LnEDU - Education	Lned_s1	1990	% of 12 and older population with secondary, logarithm	Municipality	
AGR – Agricultural	D_Agri	1990	0 if non-agricultural mun., 1 otherwise	Municipality	2,443
EM – Emigration	MIG_TOT	2000	% of 12 and older pop. outside birth municipality	Municipality	2,443

VARIABLE	DATA CODE	PERIOD	UNIT	GEOGRAPHIC AGGREGATION	OBSERVATIONS
ΔLIP – Change in Low Incomes Population	Dif_2rms	1990/2000	Difference in % of pop. below 2 minimum salaries	Municipality	2,443
LnCRE – Credit	Ln_cred	1990	Agricultural units with credit access (%), logarithm	Municipality	2,443
LnNDR – Disaster reports	Ln_reps1	1990-2000	Number of disasters reports, logarithm	Municipality	2,443
GRAVE – Grains/Vegetables	Grave00	2000	Volume ratio (tn)	State (set to municipalities)	32 (2,443)

Table E.2 List of municipalities selected as main trading centers cities

Acapulco de Juárez	Monterrey
Aguascalientes	Morelia
Ahome	Nogales
Benito Juárez	Nuevo Laredo
Cajeme	Oaxaca de Juárez
Campeche	Paz, La
Celaya	Piedras Negras
Centro	Puebla
Chihuahua	Querétaro
Coatzacoalcos	Reynosa
Cuernavaca	Salina Cruz
Culiacán	Saltillo
Durango	San Luis Potosí
Guadalajara	Tampico
Guaymas	Tapachula
Hermosillo	Tepic
Hidalgo del Parral	Tijuana
Irapuato	Toluca
Iztapalapa	Torreón
Juárez	Tuxtla Gutiérrez
León	Veracruz
Manzanillo	Victoria
Matamoros	Xalapa
Mazatlán	Zacatecas
Mérida	Zapotlán del Rey

Appendix F: List of interviewees presented in Chapter 6

Leaders, authorities and experts

2004-2005

Mr. Marco Antonio Suárez Vargas
CIOAC - Independent Union of Farmers and Peasants Organizations (Central Independiente de Organizaciones de Agricultores y Campesinos)
Regional Delegate (Region Coast-Soconusco)

Mr. Juan Pérez Pérez
CIOAC - Independent Union of Farmers and Peasants Organizations (Central Independiente de Organizaciones de Agricultores y Campesinos)
State Chair (Chiapas).

Mr. Gilberto Santos Galicia
CNC - National Confederation of Farmers (Confederación Nacional Campesina)
Local farmers' leader – Unión de Ejidos Emiliano Zapata

Ing. Indalecio Flores B.
CNC - National Confederation of Farmers (Confederación Nacional Campesina)
Regional Coordinator (Soconusco) - Liga de Comunidades Agrarias y Sindicatos Campesinos del Estado de Chiapas.

Lic. Celso Humberto Delgado
CNC - National Confederation of Farmers (Confederación Nacional Campesina)
National General Secretary

Mr. Pablo Velasco Alfaro
OPEZ - Organización Proletaria Emiliano Zapata (Proletarian Organization Emiliano Zapata)
Political Direction at the Council of the State of Chiapas

© The Author 2015
S.O. Saldaña-Zorrilla, *Natural Disasters, Foreign Trade and Agriculture in Mexico*, SpringerBriefs in Environmental Science, DOI 10.1007/978-3-319-17359-7

MVZ. Henar Galicia Negrete
SDR - Ministry of Rural Development, Government of the State of Chiapas
 (Secretaría de Desarrollo Rural)
Region Director of Rural Development in Soconusco at the

Ing. Juan Ignacio Pliego Tamayo
CDFCC - Comission for Development and Promotion of Coffee in Chiapas
 (Comisión para el Desarrollo y Fomento del Café en Chiapas)

Ing. Victor Alfonso
Municipality of Cacahoatán
Director of Agriculture

Lic. Josías Magdiel Santos Jiménez
Municipality of Cacahoatán
Municipal Judge

Lic. Jorge Antonio Saldaña Ríos
Municipality of Cacahoatán
Municipal Trustee

Ing. Gabriel de Jesús Álvarez Guzmán
Municipality of Escuintla
Deputy Director of Agricultural Encouragement

Mr. Salvador Vázquez Cárdenas
Sheriff of Ejido Tapachula

Mr. Porfirio Gutiérrez Verdugo
Sheriff of Ejido Cintalapa

Mr. Wilmar Sánchez
Deputy Sheriff of Ejido Cintalapa

Profr. Luís Guadalupe Morales Ángeles
Local leader in Ejido Cintalapa

Mr. Ricardo Gutierrez Morales
Leader of independent farmers in Cintalapa

2014

Dr. Eduardo Espinosa Medinilla
UNICACH - Universidad de Ciencias y Artes de Chiapas
Professor-Researcher

Dr. Nathan Black
Harvard University
Center for the Environment (HUCE)

Mr. Romeo García Cortés
Secretary General of NGO Escuintla Nuestro, A.C.

Prof. Rebeca Rangel
Professor in Escuintla

Mr. Wilmar Vázquez Espinos
Traders' leader in Cintalapa

Interviewed policymakers

2003

Ing. Antonio Ruiz-Garcia
SAGARPA – Ministry of Agriculture, Livestock, Rural Vedelopment, Fisheries and
  Food (Secretaría de Agricultura, Ganadería, Desarrollo Rural, Pesca y
  Alimentación)
Deputy Minister for Rural Development

Ing. Victor M. Celaya del Toro
SAGARPA – Ministry of Agriculture, Livestock, Rural Vedelopment, Fisheries and
  Food (Secretaría de Agricultura, Ganadería, Desarrollo Rural, Pesca y
  Alimentación)
Director General of Rural Development Studies

Lic. Salvador Mayoral
AGROASEMEX – State-owned Crop Insurance Company
Director General

Lic. Luis Francisco Mejia Piña
SEDESOL – Ministry of Social Development (Secretaría de Desarrollo Social)
National Coordinator of Social Programs

Lic. Hugo González Gutiérrez
FONDEN – Fund for Natural Disasters (Fondo de Desastres Naturales)
Director General

Lic. Ignacio Bermeo Juárez
FONDEN – Fund for Natural Disasters (Fondo de Desastres Naturales)
Legal and Operation Director

Ing. Roberto Quaas-Weppen
CENAPRED – National Center for Disasters Prevention (Centro Nacional de
  Prevención de Desastres) at the Ministry of the Interior.
Director General

2014

Sen. Zoe Robledo Aburto
Senator (Chiapas)

CPC Alberto José Canseco García
AGROASEMEX – State-owned Crop Insurance Company
Director General

Biol. Juan Carlos Castro Hernández
CONANP - National Commission of Natural Protected Areas
Director of the Natural Protected Area "El Triunfo"

Biol. Ricardo Hernández
Deputy Minister for the Environment
Government of the State of Chiapas

Ing. Raúl Saavedra Horita
Mexican National Committee of the International Hydrologic Program
National Affairs Secretary

Appendix G: Questionnaire to small-scale farmers presented in Chapter 6

Region: South of Mexico.
State: Chiapas.
Municipalities: Escuintla and Cacahoatán.
Communities: Escuintla, Cacahoatán, and Cintalapa
Main regional crops: coffee, cocoa, banana, maize, bean and sorghum.
Date: December 1-22, 2004; January 4-25, 2005

1. GENERAL INFORMATION

Sex: M_____ F_____
Age: _____
Property of your land: Own_____ Rented_____ Another_____
Tenure of your land: Private_____ Communal_____ *Ejidal*_____
Another_____
Access to irrigation: Yes_____ No_____
Access to credit: Yes_____ No_____
Land size (hectares):_____

1. Please, list your produced products (corn, poultry, etc.) by importance to your incomes

1. _____
2. _____
3. _____
4. _____
5. _____

2. What share of your income arises from agriculture?
   ◆ All
   ◆ Most

© The Author 2015
S.O. Saldaña-Zorrilla, *Natural Disasters, Foreign Trade and Agriculture in Mexico*,
SpringerBriefs in Environmental Science, DOI 10.1007/978-3-319-17359-7

- ◆ Half
- ◆ Less than a half
- ◆ Just a small part

3. From which non-agricultural activity do you obtain additional incomes?

4. What do you consider to be the main obstacle to generate incomes in your community?

2. NATURAL DISASTERS

5. In your point of view, what type of disaster has been the most affecting to your assets and incomes generation?

6. In the last 10 years, how many times have you been affected by natural disasters?

7. In a 1-to-5 scale, please, score the following financing sources according to how important have they been to you following a natural disaster?

Scale: 1=irrelevant, 2 = a bit important, 3 = useful, 4 = important, 5 = very important

SOURCE	SCORE (1-5)
Neighbours help	
Relatives help living:	
a) in your community	
b) somewhere else in Mexico	
c) abroad	
(Where do they reside? _____)	
Humanitarian aid:	
a) from the government	
b) private (associations, donations, etc.)	
Insurance:	
a) private	
b) public	
c) subsidized private	
d) insurance fond	
Governmental fund for property reconstruction	
Loans:	
1. Private	
2. Public	
3. Communal	
Property sell:	
a) Animals, vehicles, tools, etc.	
b) Land	
Another	

8. How are you expected to finance the recovery costs from the next natural disaster?

9. Does your community have a contingent fund to face disasters? Yes___ No___

If so, how does it work?

10. Are you currently covered by crop insurance? Yes ___ No___
If so, with who did you contracted it?

1. Private insurance (without subsidy)
2. Private insurance (with subsidy)
3. Public insurance
4. Another_____

11. What share (%) of your total cultivated area is currently insured?

12. Are there risk-reducing measures in your community so that disaster losses distribute differently upon assets and production (i.e. crops diversification)?

13. Can you obtain credit from stores and shops due to a disaster year?

14. In your opinion, who should be responsible for reducing vulnerability to natural disasters?
(Sort importance order from 1 (the most important) to 6 (the less)
____ Humanitarian aid
____ Aid from the federal government
____ Aid from the state government
____ Private insurance companies
____ Non-profitable insurance funds
____ Another (please, specify) _____

15. Who do you think should be responsible for providing financing for the private sector recovery: households and business or the government? Please, detail your answer.

3. FOREIGN TRADE

16. Have you ever exported? Yes _____ No _____
If so, has it been directly or indirectly? _____ Where to? _____
17. Approximately, which percentage of the total value of your production is exported?

18. Are you member of any exports cooperative?
Yes _____ No _____ If so, of which one?

19. What is the average freight cost for exporting your product (i.e. pesos per container, ton, trailer, etc.)?

20. From which country come the imports which compete with your products (sort by importance)?

1) _____

2) _____

3) _____

21. In your view, how has impacted the North America Free Trade Agreement (NAFTA) and further trade agreements, first, to you, and also to Mexico?

22. Do you have plans to diversify your crops? Yes _____ No _____ If so, what are you planning to produce (additionally)? _____ _____ _____

23. Which crops are you planning to leave? _____ _____ _____

24. Have you ever obtained governmental advice concerning crops diversification? Yes _____ No _____ By means of which agency? _____

25. What would you suggest for improving agricultural trade conditions in your region?

4. MIGRATION

26. Have you ever imagined a situation where you leave your community (i.e. if agricultural prices keep dropping, natural disasters, etc.)?

27. Where would you go to?

28. Have any of your relatives been forced to emigrate or to change activity due to natural disasters? If so, where have they emigrated to or which activity have they shifted to?

29. Do you have relatives residing abroad? Yes ____ No____
If so, where? _____

30. Currently, do you have emigration plans? Yes ____ No ____
If so, to what type of community? rural ___ urban ____
Where? In Mexico____ abroad _____ Where exactly? _____

31. If any, what is the cause of your will to emigrate?

32. Given current conditions, would you encourage a relative to emigrate?

5. FINAL QUESTIONS

33. Do you consider governmental support to be really reaching its target population and to achieve its goals? Yes ____ No____ Why?

34. What would you ask the government to implement in your community to reduce vulnerability to natural disasters?

35. What would you ask the government to implement in your community to reduce vulnerability to agricultural uncertainty?

PERSONAL COMMENTS

Index

© The Author 2015
S.O. Saldaña-Zorrilla, *Natural Disasters, Foreign Trade and Agriculture in Mexico*,
SpringerBriefs in Environmental Science, DOI 10.1007/978-3-319-17359-7

Printed in the United States
By Bookmasters